T0185638

Nutrigenomics: How Science Works

Carsten Carlberg • Stine Marie Ulven
Ferdinand Molnár

Nutrigenomics: How Science Works

 Springer

Carsten Carlberg
Institute of Biomedicine
University of Eastern Finland
Kuopio, Finland

Stine Marie Ulven
Department of Nutrition
University of Oslo
Oslo, Norway

Ferdinand Molnár
Department of Biology
Nazarbayev University
Nur-Sultan, Kazakhstan

ISBN 978-3-030-47663-2 (PB)
ISBN 978-3-030-36947-7 (HB)   ISBN 978-3-030-36948-4   (eBook)
https://doi.org/10.1007/978-3-030-36948-4

This Springer imprint is published by the registered company Springer Nature Switzerland AG.
The registered company address is: Gewerbestrasse 11, 6330 Cham, Switzerland

Preface

Our daily diet is more than a collection of carbohydrates, lipids, and proteins, minerals, and vitamins that provide energy and serve as building blocks of our life. It is also the most dominant environmental signal to which we are exposed from womb to death. The availability of the sequence of the complete human genome and the consequent development of next-generation sequencing technologies have significantly affected nearly all areas of bioscience. This was the starting point for new disciplines, such as genomics and its subdiscipline nutrigenomics. The fascinating area of **nutrigenomics describes the daily communication between dietary molecules, their metabolites, and our genome**. Its genomic components comprise not only the variation of the human genome, such as SNPs (single-nucleotide polymorphisms), but also the dynamic packaging of the genome into chromatin, including all information stored in this epigenome. Moreover, this book discusses the proteins that are involved in the signal transduction between dietary molecules and the genome, such as nuclear receptors, chromatin-modifying enzymes, and energy status-sensing kinases, and their mechanism of action.

Most noncommunicable diseases, such as T2D (type 2 diabetes) and CVDs (cardiovascular diseases), are the basis of lifestyle decisions. They have not only a genetic, inherited component, but to some 80%, they are based on epigenetics (meaning "above" genetics), i.e., on our **lifestyle choices and environmental exposures, such as what we eat**. We cannot change the genes that we are born with, but we can take care of the rest being primarily based on our epigenome. This means that the genetic predisposition for a disease can be counterbalanced by an appropriate healthy lifestyle that modulates the epigenome of the affected tissues. It is well known that there is a high level of individual responsibility for staying healthy, but a detailed understanding of epigenetics provides a molecular explanation for this life attitude. This book describes how nutrition shapes human evolution and demonstrates its consequences for our susceptibility to diseases, such as T2D and atherosclerosis, the underlying cause of most CVDs. Inappropriate diet can yield stress for our cells, tissues, and organs, and then it is often associated with low-grade chronic inflammation. Overnutrition paired with physical inactivity leads to overweight and obesity and results in increased burden for a body that originally was adapted for a life in the savannas of East Africa. Thus, this book does not only discuss about a theoretical topic in science, but it especially talks about real life and our lifelong "chat" with diet. **We are all food consumers; thus, each of us is concerned by the topic of this book and should be aware of its mechanisms of how our key daily lifestyle decision affects our health**.

The purpose of this book is to provide an overview on the **principles of nutrigenomics and their relation to health and disease**. We are not aiming to compete with more comprehensive textbooks on molecular nutrition, evolutionary biology, genomics, gene regulation, or metabolic diseases but rather will focus on the essentials and will combine, in a compact form, elements from different disciplines. In order to facilitate the latter, we favor a high figure-to-text ratio following the rule "a picture tells more than thousand words."

The content of the book is linked to a series of lecture courses in "Molecular Medicine and Genetics," "Molecular Immunology," "Cancer Biology," and "Nutrigenomics" that are given by one of us (C. Carlberg) in different forms since 2002 at the University of Eastern Finland in Kuopio. This book represents an updated version of our textbook *Nutrigenomics* (ISBN 978-3-319-30415-1). However, we shortened and simplified the content in order to give also undergraduate students and other people engaged in life sciences an easier start into the topic. This book also relates to our textbooks *Mechanisms of Gene Regulation* (ISBN 978-94-017-7741-4), *Human Epigenomics* (ISBN 978-981-10-7614-8), and *Human Epigenetics: How Science Works* (ISBN 978-3-030-22906-1), the studying of which may be interesting to readers who like to get more detailed information. Following two introductory chapters, the first five chapters of this book will explain the molecular basis of nutrigenomics, while the last three chapters will provide examples for the impact of nutrigenomics on our health and disease. A glossary in the appendix will explain the major specialist's terms.

We hope that the readers will enjoy this rather visual book and get as enthusiastic about nutrigenomics as the authors are.

Kuopio, Finland   Carsten Carlberg
Oslo, Norway   Stine Marie Ulven
Nur-Sultan, Kazakhstan   Ferdinand Molnár
October 2019

Acknowledgments

The authors would like to thank Eunike Velleuer, MD, and Andrea Hanel, BSc, for extensive proofreading and constructive criticism.

Contents

Abbreviations

1,25(OH)$_2$D$_3$	1,25-dihydroxyvitamin D$_3$
25(OH)D$_3$	25-hydroxyvitamin D$_3$
3D	3-dimensional
α-MSH	α-melanocyte-stimulating hormone
ABC	ATP-binding cassette
ABL	abetalipoproteinemia
AC	adenylate cyclase
ACAT1	acetyl-CoA acetyltransferase 1
ACC	acetyl-CoA carboxylase
ACL	ATP citrate lyase
ADAMTS	ADAM metallopeptidase with thrombospondin motif
ADH	alcohol dehydrogenase
ADP	adenosine diphosphate
ADRB3	adrenoceptor beta 3
AGRP	agouti-related peptide
AKT	AKT serine/threonine kinase
ALOX5	arachidonate 5-lipoxygenase
ALOX15	arachidonate 15-lipoxygenase
AMPK	adenosine monophosphate-activated protein kinase
AMY	amylase
ANGPTL2	angiopoietin-like protein 2
APEH	N-acylaminoacyl-peptide hydrolase
AP1	activating protein 1
APO	apolipoprotein
APPL1	adaptor protein, phosphotyrosine interacting with PH domain and leucine zipper 1
AR	androgen receptor
ARC	arcuate nucleus
ARID	AT-rich interaction domain
ARL4C	ADP-ribosylation factor-like 4C
ARNTL	aryl hydrocarbon receptor nuclear translocator-like

ASC   apoptosis-associated speck
ASIP   agouti-signaling protein
ATF6   activating transcription factor 6
ATP   adenosine triphosphate
β-OHB   β-hydroxybutyrate
BAAT   bile acid-CoA:amino acid N-acetyltransferase
BAT   brown adipose tissue
BDNF   brain-derived neurotrophic factor
BLK   B lymphoid tyrosine kinase
BMI   body mass index
BMP   bone morphogenetic protein
bp   base pair
BRD   bromodomain containing
CAMKK Ca^{2+}/calmodulin-dependent protein kinase kinase
CAMP   cathelicidin
CAR   constitutive androstane receptor
CASP   caspase
CBL   Cbl proto-oncogene, E3 ubiquitin protein ligase
CCK   cholecystokinin
CCL   chemokine (C-C motif) ligand
CCR   C-C chemokine receptor
CD36   CD36 molecule
CDC42   cell division cycle 42
CDKAL1   CDK5 regulatory subunit associated protein 1-like 1
CDKN   cyclin-dependent kinase inhibitor
CDP   common dendritic cell progenitor
CDX2   caudal type homeobox 2
CEBP   CCAAT-binding protein
CEL   carboxyl ester lipase
CETP   cholesterol ester transfer protein
CETPD   CETP deficiency
CHD   coronary heart disease
CHGA   chromogranin A
ChIP   chromatin immunoprecipitation
CLOCK   clock circadian regulator
CLP   common lymphoid progenitor
CMP   common myeloid progenitor
CNS   central nervous system
CNV   copy number variant
CPT1A   carnitine palmitoyltransferase 1A
CREB3L3   cAMP responsive element binding protein 3-like 3
CREBBP   CREB-binding protein, also called KAT3A
CRP   C-reactive protein
CRTC2   CREB-regulated transcription coactivator 2
CRY1   cryptochrome circadian clock 1

CSF2	colony-stimulating factor 2
CTNS	cystinosin, lysosomal cystine transporter
CVD	cardiovascular disease
CXCL5	chemokine (C-X-C motif) ligand 5
CXCR	C-X-C motif chemokine receptor
CYP	cytochrome P450
DAF	abnormal dauer formation
DAG	diacylglycerol
DALY	disability-adjusted life year
DAMP	damage-associated molecular pattern
DBL	dysbetalipoproteinemia
DC	dendritic cell
DCT	dopachrome tautomerase
DEFB4	defensin, beta 4A
DNMT	DNA methyltransferase
DOHaD	Developmental Origins of Health and Disease
E%	percent of total energy
EDAR	ectodysplasin A receptor
EIF2A	eukaryotic translation initiation factor 2A
EIF2AK3	eukaryotic translation initiation factor 2-alpha kinase 3
EGIR	European Group for the Study of Insulin Resistance
EHMT	euchromatic histone-lysine N-methyltransferase
ENCODE	Encyclopedia of DNA Elements
ENPP1	ectonucleotide pyrophosphatase/phosphodiesterase 1
EP300	E1A binding protein p300, also called KAT3B
eQTL	expression quantitative trait locus
ER	endoplasmic reticulum
ERN1	endoplasmic reticulum to nucleus signaling 1
FABP6	ileal fatty acid binding protein
FAD	flavin adenine dinucleotide
FANTOM	functional annotation of the mammalian genome
FAS	Fas cell surface death receptor
FASN	fatty acid synthase
FCH	familial combined hyperlipidemia
FFA	free fatty acid
FGF	fibroblast growth factor
FGFR4	FGF receptor 4
FH	familial hypercholesterolemia
FHC	familial hyperchylomicronemia
FHTG	familial hypertriglyceridemia
FOXO	forkhead box O
FTO	fat mass and obesity-associated
FXR	farnesoid X receptor
G6PC	glucose-6-phosphatase
GAB1	GRB2-associated binder 1

GATA	GATA binding protein
GCK	glucokinase
GC	gas chromatography
GH1	growth hormone 1
GIS1	GIg1-2 suppressor
GLP1	glucagon-like peptide 1
GMP	granulocyte-monocyte progenitor
GPAT	glycerol-3-phosphate acyltransferase
GPR	G-protein-coupled receptor
GR	glucocorticoid receptor
GRB	growth factor receptor-bound protein
GSK3	glycogen synthase kinase 3
GSV	GLUT4-containing storage vesicles
GWAS	genome-wide association study
GYS	glycogen synthase
HAT	histone acetyltransferase
HBL	hypobetalipoproteinemia
HDAC	histone deacetylase
HDM	histone demethylase
HDL	high-density lipoprotein
HHEX	hematopoietically expressed homeobox
HIF1	hypoxia-inducible factor 1
HIV	human immunodeficiency virus
HLA	human leukocyte antigen
HLD	hepatic lipase deficiency
HLP	hyperlipoproteinemia
HMGCR	3-hydroxy-3-methylglutaryl-CoA reductase
HMT	histone methyltransferase
HNF	hepatocyte nuclear factor
HPT	hypothalamic-pituitary-thyroid
HSC	hematopoietic stem cell
HSF1	heat shock transcription factor 1
HSP	heat shock protein
HTG	hypertriglycerolemia
IAP	intracisternal A particle
IDF	International Diabetes Federation
IDH	isocitrate dehydrogenase
IDOL	inducible degrader of LDLR
IGF	insulin-like growth factor
IGF1R	IGF1 receptor
IGF2BP2	insulin-like growth factor 2 mRNA binding protein 2
IKBK	inhibitor of kappa light polypeptide gene enhancer in B cells, kinase
IL	interleukin
IL1RN	IL1 receptor antagonist
IMCL	intramyocellular lipid

indel	insertion-deletion
INFG	interferon γ
INS	insulin
iPOP	integrative personal omics profiling
IR	insulin receptor
IRE1	inositol-requiring enzyme
IRF	interferon regulatory factor
IRS	insulin receptor substrate
IRX	iroquois homeobox
JAK	Janus kinase
K^{ATP}	ATP-sensitive K^+
kb	kilo base pairs (1000 bp)
KCNJ11	potassium inwardly rectifying channel, subfamily J, member 11
KCNQ1	potassium voltage-gated channel subfamily Q, member 1
KDM	lysine demethylase
KLF	Krüppel-like factor
KMT	lysine methyltransferase
LCAT	lecithin cholesterol acyltransferase
LCATD	LCAT deficiency
LCR	locus control region
LCT	lactase
LDL	low-density lipoprotein
LDLR	LDL receptor
LDLRAP1	LDLR accessory protein 1
LEP	leptin
LEPR	leptin receptor
LINE	long interspersed element
LIPC	hepatic lipase
LIPE	hormone-sensitive lipase
LIPG	endothelial lipase
LPCAT3	lysophospholipid acyltransferase 3
LPL	lipoprotein lipase
LRH-1	liver receptor homolog 1
LRP1	LDLR-related protein 1
LTR	long terminal repeat
LXR	liver X receptor
MAF	minor allele frequency
MAFA	v-maf avian musculoaponeurotic fibrosarcoma oncogene homolog A
MAN2A1	mannosidase, alpha, class 2A, member 1
MAPK	mitogen-activated protein kinase
Mb	mega base pairs (1,000,000 bp)
MC1R	melanocortin 1 receptor
MC4R	melanocortin 4 receptor
M-CFU	myeloid stem cells

MCM6	minichromosome maintenance type 6
MDP	macrophage and dendritic cell progenitor
MHC	major histocompatibility complex
MHL	mixed hyperlipidemia
mmHg	millimeters of mercury
MODY	maturity onset diabetes of the young
MPO	myeloperoxidase
MPP	multipotent progenitor
MS	mass spectrometry
MSN	multicopy suppressor of SNF1 mutation
MSR1	macrophage scavenger receptor 1
MTHFR	methylenetetrahydrofolate reductase
MTNR1B	melatonin receptor 1B
MTTP	microsomal triglyceride transfer protein
MUFA	monounsaturated
MYCL	v-myc avian myelocytomatosis viral oncogene lung carcinoma derived homolog
MYF5	myogenic factor 5
MYO5A	myosin VA
NAD	nicotinamide adenine dinucleotide
NAFLD	nonalcoholic fatty liver disease
NAMPT	nicotinamide mononucleotide phosphoribosyltransferase, also called visfatin
NANOG	nanog homeobox
NCOA	nuclear receptor coactivator
NCEH1	neutral cholesterol ester hydrolase 1
NCEP	National Cholesterol Education Program
ncRNA	noncoding RNA
NEUROD1	neuronal differentiation 1
NEUROG3	neurogenin 3
NFκB	nuclear factor κB
NLR	NOD-like receptor
NLRP	NLR protein
NO	nitric oxide
NOS2	inducible nitric oxide synthase 2
NPC1L1	Niemann-Pick C1-like protein 1
NPY	neuropeptide Y
NTS	nucleus tractus solitarius
OCA2	OCA2 melanosomal transmembrane protein
OGTT	oral glucose tolerance test
ORL1	oxidized low-density lipoprotein receptor 1
PAMP	pathogen-associated molecular pattern
PAX	paired box
PBMC	peripheral blood mononuclear cell
PC	pyruvate carboxylase

PCK	phosphoenolpyruvate carboxykinase
PCSK1	proprotein convertase subtilisin/kexin type 1
PDH	pyruvate dehydrogenase
PDPK	3-phosphoinositide dependent protein kinase 1
PDX1	pancreatic and duodenal homeobox 1
PER1	period circadian clock 1
PFKFB2	6-phosphofructo-2-kinase/fructose-2,6-biphosphatase 2
PI3K	phosphoinositide 3-kinase
PIP3	phosphatidylinositol-3,4,5-triphosphate
PKA	protein kinase A
PLAU	plasminogen activator, urokinase
PLTP	phospholipid transfer protein
PNPLA	patatin-like phospholipase domain-containing protein-3
Pol II	RNA polymerase II
POMC	proopiomelanocortin
POU1F1	POU class 1 homeobox 1
POU5F1	POU class 5 homeobox 1
PPAR	peroxisome proliferator-activated receptor
PPARGC1A	PPAR gamma, coactivator 1 alpha
PPP2	protein phosphatase 2
PRK	protein kinase
PROP1	PROP paired-like homeobox 1
PRR	pattern recognition receptor
PUFA	polyunsaturated fatty acid
PTEN	phosphatase and tensin homolog
PTGS2	prostaglandin-endoperoxide synthase 2 (also known as COX2)
PTPN1	protein tyrosine phosphatase, non-receptor type 1
PXR	pregnane X receptor
RAPTOR	regulatory-associated protein of TOR
RAR	retinoic acid receptor
RBP4	retinol binding protein 4
RE	response element
REV-ERB	Reverse-Erb, official gene symbol NR1D1
RHOQ	Ras homolog family, member Q
RIG1	retinoic acid-inducible gene 1
RLR	RIG1-like helicase receptors
RNA-seq	RNA sequencing
ROR	RAR-related orphan receptor
ROS	reactive oxygen species
RPS6K	ribosomal protein S6 kinase
RXR	retinoid X receptor
S6K	S6 kinase
SAH	S-adenosylhomocysteine
SAM	S-adenosylmethionine
SCAP	SREBF chaperone

SCARB1	scavenger receptor class B member 1
SCD1	steroyl-CoA desaturase 1
SCN	suprachiasmatic nucleus
SCNN1	sodium channel, non-voltage-gated 1
SERPINE1	serpin peptidase inhibitor, clade E (also called PAI-1)
SF-1	steroidogenic factor 1
SFA	saturated fatty acids
SFRP5	frizzled-related protein 5
SH2	Src homology 2
SHC	Src homology 2 domain-containing
SHP2	SH2-domain-containing tyrosine phosphatase 2
SI	sucrase-isomaltase
SIM1	single-minded family bHLH transcription factor 1
SINE	short interspersed element
SIRT	sirtuin
SITO	sitosterolemia
SLC	solute carrier family
SLCO	solute organic anion transporter
SNP	single-nucleotide polymorphism
SNS	sympathetic nervous system
SOCS	suppressor of cytokine signaling
SORBS1	sorbin and SH3 domain-containing 1
SOS	Son of Sevenless
SORT1	sortilin 1
SPI1	spleen focus-forming virus proviral integration oncogene, also called PU.1
SREBF1	sterol regulatory element-binding transcription factor 1
STAT	signal transducer and activator of transcription
SULT2A1	sulfotransferase family 2A, member 1
T1D	type 1 diabetes
T2D	type 2 diabetes
TAS1R2	taste receptor, type 1, member 2
TBC1D	TBC1 domain family, member 1
TCA	tricarboxylic acid
TCGA	The Cancer Genome Atlas
TD	Tangier disease
TET	ten-eleven translocation
TGFB1	transforming growth factor beta 1
T_H	T helper
THRSP	thyroid hormone responsive
TLR	Toll-like receptor
TNF	tumor necrosis factor
TNFR	TNF receptor
TOR(C)	target of rapamycin (complex)
TRAF2	TNF receptor-associated factor 2

T_{REG}	regulatory T
TSC2	tuberous sclerosis 2
TSS	transcription start site
TYR	tyrosinase
UBR1	ubiquitin protein ligase E3 component n-recognin 1
UCP	uncoupling protein
UGT2B4	UDP glucuronosyltransferase 2 family, polypeptide B4
UNC5B	unc-5 homolog B
UV	ultraviolet
VDR	vitamin D receptor
VLDL	very low-density lipoprotein
VNN	vanin 1
WAT	white adipose tissue
WHO	World Health Organization
WHR	waist-hip ratio
WNT	wingless-type MMTV integration site family member
YWHA	tyrosine 3-monooxygenase/tryptophan 5-monooxygenase activation protein (also called 14-3-3)
XBP1	X-box binding protein 1

Chapter 1
Nutrition and Common Diseases

Abstract This chapter will provide a first overview of the role of nutrition on our health and disease. During the past 50 years a significant lifestyle change has happened to nearly all humans worldwide. This is characterized by the use of energy-rich, highly processed food paired with reduced physical activity. Diet is one of the key environmental factors particularly involved in the pathogenesis and progression of many chronic non-communicable diseases, such as obesity, T2D and CVDs like hypertension, myocardial infarction and stroke as well as cancer. We will describe the evidence of dietary factors for these diseases and the impact of physical activity on their prevention. Obesity and cancer and will serve as examples, in order to describe the link between inflammation and nutrition-triggered diseases.

Keywords Nutrition · Evolution · Catabolism · Anabolism · Non-communicable diseases · Obesity · Cancer · Physical activity · Inflammation

1.1 Evolution of Human Nutrition

Nutrition is essential for life, but the effects of nutritional molecules on our health are complex and influenced by many factors. Our diet is composed of food groups that collectively provide our body with its nutritional needs of macro- and micronutrients (Sect. 1.3). In addition to nutrients, food also contains hundreds of bioactive compounds that have an effect on our metabolism. Single nutrients or food groups have relatively small effects on our health, but **the overall quality of diet and the interaction among many nutrients is critical**. In this context the impact of healthy dietary patterns, such as Mediterranean or Nordic diet (Box 1.1), needs to be understood.

   The main theme of this book is the daily communication between our diet and our genome, which modulates gene regulatory networks in our metabolic organs, such as in skeletal muscle, adipose tissue, pancreas and liver, as well as in our immune system and brain. The molecular and cellular processes controlled by these networks keep our body in homeostasis and prevent the onset of non-

© Springer Nature Switzerland AG 2020
C. Carlberg et al., *Nutrigenomics: How Science Works*,
https://doi.org/10.1007/978-3-030-36948-4_1

Box 1.1 Mediterranean and Nordic diet

Mediterranean diet is based on the eating habits of Greece and Italy in the 1960s. It includes rather high amounts of olive oil, unrefined cereals, fruits and vegetables, moderate consumption of fish and dairy products, such as cheese, and yogurt, and red wine as well as low consumption of non-fish meat products. Mediterranean diet is associated with a reduction in mortality, primarily by lowering the risk of heart disease and T2D. Nordic diet highlights the local, seasonal and nutritious foods from Denmark, Norway, Sweden, Finland and Iceland. Like Mediterranean diet it is high in plant-based foods, *i.e.*, it emphasizes whole grains, such as barley, rye and oats, berries, vegetables and fatty fish as well as low-fat dairy and low consumption of red meat. However, instead of olive oil, the Nordic diet is rich in rapeseed oil.

communicable diseases, such as obesity, T2D, CVDs and cancer. Thus, **an appropriate diet in combination with sufficient physical activity** (Sect. 1.6) **will keep us healthy**.

Anatomically modern humans (*Homo sapiens*) evolved some 300,000 years ago in Africa and started some 60–80,000 years ago to spread in larger numbers over the whole planet (Sect. 2.1). Until some 10,000 years ago they lived as hunters and gatherers, *i.e.*, their diet was based on wild animals and plants (Table 1.1). Depending on season and geographic region they were eating meat, fish, eggs, fruits, nuts and seeds. Accordingly, food was of rather low energy density, *i.e.*, it provided fewer calories per gram, had medium fiber content, was rich in starch and low in fat, had rather high micronutrient density and anti-oxidant capacity and was low in salt. Thus, **the biochemistry of our body had the time span of some 300,000 years, *i.e.*, approximately 12,000 generations, to adapt to this food**.

Agricultural revolution, *i.e.*, the use of domesticated plants and animals, started some 10,000 years ago and shifted the human diet pattern to even higher rates in starch, the introduction of sugar and the use of salt for conservation. This increased the energy density of food but also its fiber content and resulted in a high salt load. However, relative to the diets of many hunter-gatherer societies the agricultural transition resulted in reduced nutritional diversity, since starch alone already represented more than 60% of the daily caloric intake.

The industrial revolution over the past 250 years, such as the use of machines, trains and cars, reduced the need of physical activity for work and transport. In parallel, the use of refined food, such as grains and oils, increased. This made diet lower in starch and fiber but enriched in white sugar and fat, leading to higher energy density and glycemic load. Finally, in modern times we primarily use industrially processed dietary products and so-called "fast food". In particular, the latter is of very high sugar and fat content, *i.e.*, of high energy density but low fiber content. In parallel, technical revolutions in transportation and computerization further

Table 1.1 Evolution of human nutrition. Change of human diet from the Paleolithic era *via* agricultural and industrial revolution to modern times is indicated

Time period	Diet	Nutritional characteristics
Paleolithic era (more than 10,000 years ago)	Wild animals and plants, varied by geography and season, such as meat, fish, eggs, fruits, nuts and seeds	Low energy density
		Medium fiber content (40 g/day)
		Macronutrients: 15–20% protein, 50–70% starch and 15–20% fat
		Low glycemic load
		Low salt (1 g/day): Na/K ratio < 1
Agricultural revolution (starting 10,000 years ago)	Largely based on domesticated animals and plants, such as grains, dairy products, vegetables. Use of fermented foods and beverages	Medium energy density
		High fiber content (60–120 g/day)
		Macronutrients: 10–15% protein, 60–75% starch, 5% sugar and 10–15% fat
		High glycemic load
		High salt (5–15 g/day): Higher Na/K ratio
Industrial revolution (starting 250 years ago)	Increased reliance of refined grains and oils, fatty meat, alcoholic beverages	High energy density
		Medium fiber content (40 g/day)
		Macronutrients: 12% protein, 40–50% starch, 10% sugar and > 30% fat
		High glycemic load
		High salt (10 g/day): Higher Na/K ratio
Modern era (starting 50 years ago)	Mainly industrially produced foods, such as refined grains and oils, fatty meat from domesticated animals, alcoholic beverages.	Very high energy density
		Low fiber content (20 g/day)
		Macronutrients: 15% protein, 25% starch, > 20% sugar and > 40% fat
	Consumption of highly processed "fast food"	Very high glycemic load
		High salt (10 g/day): Higher Na/K ratio

reduced the need of physical activity, so that an increasing proportion of the population got a positive energy balance leading to a worldwide epidemic of overweight and obesity (Sect. 8.1). **The "energy flipping point" happened in high-income countries already some 50 years ago but nowadays reached nearly every society worldwide**. Thus, the most significant change in our nutritional habits happened in less than two generations, *i.e.*, in a time span that is far too short for expecting any genetic adaptation (Sect. 4.1).

1.2 Principles of Metabolism

Life follows the laws of thermodynamics indicating that a constant intake of energy is essential for the maintenance of highly ordered structures, such as cells, organs and whole organisms. Thus, **we have to eat, in order to keep our body intact**. Our nutrition is the provision to obtain a sufficient amount of macro- and micronutrients necessary to support essential functions of life, such as energy supply, reproduction and growth. The main macronutrients in diet are carbohydrates (digested to glucose), proteins (digested to amino acids) and lipids (dissolved into fatty acids and cholesterol) (Fig. 1.1), whereas micronutrients consist of vitamins and minerals (Chap. 3). In catabolism macronutrients provide the body with energy when they

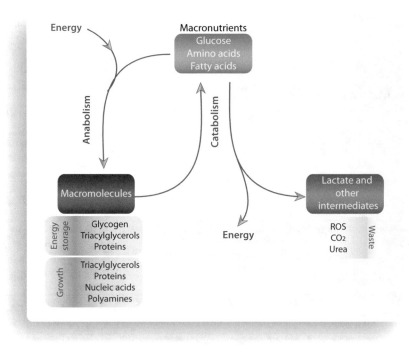

Fig. 1.1 Simplified view of metabolic principles. Macronutrients provide our body with energy *via* catabolic processes or are used for energy storage in form of glycogen (from glucose) and triacylglycerols (from fatty acids). During starvation also skeletal protein is used as a source of energy. When macronutrients are catabolized to generate energy, lactate and other intermediates, such as ROS (reactive oxygen species), carbon dioxide and ammonia, are produced. The body needs to get rid of this waste *via* endogenous anti-oxidant defense systems, *i.e.*, exhaling *via* the lungs and urea excretion *via* kidney and bladder, in order to avoid intoxication. Furthermore, our body also uses amino acids for the synthesis of nucleic acids and polyamines for normal growth and regeneration

are metabolized. When there is excess of energy-rich nutrients, they are stored in form of glycogen in liver and skeletal muscle or triacylglycerols (*i.e.*, glycerol esterified with three fatty acids) in adipose tissue. Equally, these nutritional compounds can be used in anabolism for the synthesis of new macronutrients. Both catabolism and anabolism are represented by a series of tightly controlled biochemical pathways composed of enzyme cascades, many of which require micronutrients as coenzymes (Fig. 1.1).

Food intake is based on the interaction between homeostatic regulation and hedonic sensations. During and shortly after a meal a variety of hormones and nutrients are released by the gastrointestinal tract and associated glands, and these are transported in the blood to the CNS (central nervous system). Thus, **our brain controls satiety and hunger and regulates in this way the energy balance of our whole body** (Sect. 8.4).

After a meal, the liver is the first organ to receive *via* the portal vein nutrients absorbed in the intestine. The liver plays a major role in energy storage, in particular of carbohydrates (Sect. 9.1), but it also metabolizes amino acids and fatty acids. Adipose tissue (Sect. 8.2) stores most of our body's energy as triacylglycerols, while in a process called lipolysis this organ releases fatty acids when other tissues need energy. For storing energy, fat (9 kcal/g) is more than double as efficient than carbohydrates (4 kcal/g). Thus, **we use stored fat for surviving the daily and seasonal feast-famine cycles**.

Fat powers up running endurance, which is the evolutionary basis for predatory behavior of hunters and gatherers. Skeletal muscle is a highly active organ that uses both types of stored energy, *i.e.*, glycogen and triacylglycerols, and takes up glucose and fatty acids directly from the circulation depending on the type of exercise performed (Sect. 1.6). However, the continuous food exposure and the sedentary lifestyle of modern civilization made properties, such as surviving starvation or running long distances to track down animals, dispensable. Thus, **permanent obesity became a key trait of modern humans**.

1.3   Dietary Molecules

Dietary carbohydrates consist of

- polysaccharides, such as starch
- disaccharides, such as sucrose, maltose and lactose
- free monosaccharides, such as glucose and fructose.

To maintain health and prevent non-communicable diseases, the acceptable distribution range of carbohydrates in our diet should represent 50–60 percent of total energy (E%), while added sugars should not exceed 10 E%. Glucose is the most common dietary monosaccharide and is absorbed directly into the bloodstream during digestion of polysaccharides and disaccharides. **Glucose serum concentration is strictly regulated by the peptide hormones insulin and glucagon** (Sect. 9.1), because the CNS is largely reliant on glucose as its metabolic fuel and red blood cells are even entirely dependent on it. Nutrients or their metabolites regulate the expression of genes involved in these metabolic pathways either directly or indirectly *via* insulin. Therefore, disturbances in insulin signaling (Sect. 6.3) are the major cause of the metabolic syndrome (Sect. 10.4). After digestion, glucose is taken up in all tissues by specific transport proteins and phosphorylated to glucose-6-phosphate (Sect. 3.1) before entering the pathways of glycogenesis (storage) or glycolysis (anaerobic energy production). The end product of glycolysis is pyruvate that enters the TCA (tricarboxylic acid) cycle, in order to generate energy more efficiently due to use of oxygen (oxidative phosphorylation). During overnight fasting, glycogen is broken down to glucose *via* glycogenolysis and the body starts to synthesize glucose from amino acids or other small molecules *via* gluconeogenesis, in order to ensure energy supply from sufficient concentration of glucose in the blood (5 mM).

Proteins are polymers of a set of 20 different amino acids. Nine of these amino acids are "essential", *i.e.*, our body cannot synthesize them and we must obtain them from our diet. All processes in life, ranging from control of metabolism *via* immune functions to physical movement, are dependent on these "workers" of the cell. The recommended daily intake of protein is 10–20 E%. In addition, a number of amino acids are used for other purposes than protein synthesis, such as the synthesis of the nucleic acid components purines and pyrimidines. Normally, the rate of amino acid breakdown balances the rate of their intake, *i.e.*, **our body does not store amino acids as an energy source**. However, during extreme situations, such as starvation or disease, our body breaks down proteins to amino acids, in order to use their carbon backbone as substrates for the synthesis of glucose, fatty acids and ketone bodies. Thus, amino acids can play an important role in whole body energy homeostasis.

Lipids are a major source of energy for our body with a recommended daily intake up to 35 E%. Triacylglycerols represent the majority of dietary fat, while cholesterol and phospholipids are of lower amount. Fatty acids are classified into saturated (SFAs), monounsaturated (MUFAs) or polyunsaturated (PUFAs) depending on the number of double bonds in their backbone structure. The proportion of the fatty acids in food depends on its source, such as animal (SFAs) or plant origin (MUFAs and PUFAs). The recommended daily intake of MUFAs should be in the order of 10–15 E%, that of SFAs less than 10 E% and that of PUFAs 5–10 E%. Lipids are insoluble in aqueous solutions and are therefore found in all cells associated with membranes: in adipocytes as triacylglycerol droplets and in blood plasma as major components of differently sized lipoproteins (Sect. 10.3). Dietary lipids are not only important suppliers of energy (*via* fatty acid β-oxidation), but some of them, such as selected fatty acids, steroid hormones and eicosanoids (*i.e.*, oxidized PUFAs with 20 carbon atoms), also act as co-enzymes and biological active molecules (Box 1.2) with critical roles in the control of whole body's homeostasis (Chap. 3 and 9). **Disturbances in lipid metabolism, *i.e.*, dyslipidemias** (Sect. 10.3)**, are common in chronic metabolic diseases**.

Box 1.2 Dietary components acting as signaling molecules
Biological active molecules either carry signals over long distances (endocrine signaling), act locally to transfer information between neighboring cells (paracrine signaling) or communicate within the cell itself (autocrine signaling). The lipophilic fraction of these molecules, such as steroid hormones or eicosanoids (*e.g.*, prostaglandins), can cross the plasma membrane and bind to transcription factors in the cytoplasm or nucleus (Sect. 3.2), whereas the larger hydrophilic fraction of signaling molecules binds to membrane proteins on the surface of target cells. Macronutrients, such as fatty acids, cholesterol, glucose and amino acids, and micronutrients, such as vitamin A, vitamin D, vitamin E, calcium and iron, can either act as ligands of nuclear receptors or as co-factors to enzymes. Nutrients can also bind to membrane proteins and initiate intercellular signaling pathways leading to changes in the activity of transcription factors (Sect. 4.4). Target genes regulated by transcription factors are encoding for proteins that play important roles in transport, uptake and storage of nutrients and as enzymes in metabolic pathways. Thus, these dietary components play critical roles in the control of energy homeostasis (Sect. 3.3).

1.4 Nutrition and Metabolic Diseases

Non-communicable diseases, such as cancer, CVD and T2D, cause more than 70% of early deaths worldwide and represent the leading cause of premature disability. However, **70–90% of non-communicable diseases are preventable**. A common cause of these diseases is obesity, which is defined by the WHO (World Health Organization) as excessive fat accumulation (BMI (body mass index) $\geq$ 30 kg/m^2). Obesity leads to 5–20 years decreased life expectancy of the individual (Sect. 8.1).

The *Global Burden of Disease Study* indicated that in 2017 11 million deaths and 255 million DALYs (disability-adjusted life-years) were attributable to dietary risk factors, of which the main risks were high intake of sodium (three million deaths and 70 million DALYs), low intake of whole grains (three million deaths and 82 million DALYs) and low intake of fruits (two million deaths and 65 million DALYs) (Sect. 10.1). **Fiber is a dietary factor that convincingly reduces the risk of weight gain and obesity** (Table 1.2). Moreover, regular physical activity also convincingly reduces the risk of obesity (Sect. 1.6). In contrast, the dietary factor that convincingly increases the risk of weight gain and obesity is a high intake of processed foods that are not only energy dense but also micronutrient poor (also referred as "empty calorie" foods). Typical energy dense foods are high in fat (butter, oils and fried foods), sugar or starch. In contrast, energy-dilute foods have high content of water and fiber, such as fruits, vegetables, legumes and whole grain cere-

Table 1.2 Overview of lifestyle factors and risk of developing obesity

Evidence	Decreased risk	No relationship	Increased risk
Convincing	Walking		Sugar sweetened drinks
			Screen time (children)
Probable	Regular physical activity[a]		"Western type" dietary pattern[b]
	Foods containing fiber		"Fast foods"
	"Mediterranean type" dietary pattern		Screen time (adults)
	Having been breastfed		
Possible	Low glycemic index foods	Protein content of the diet	Large portion sizes
			High proportion of food prepared outside the home (developed countries)
			"Rigid restraint/periodic disinhibition" eating patterns
Insufficient	Increased eating frequency		Alcohol

[a]Aerobic physical activity only
[b]Such diets are characterized by high intakes of free sugar, meat and dietary fat

als. Other dietary factors that increase the risk of weight gain and obesity are sugar-sweetened soft drinks and fruit juices.

Obesity leads to low-grade chronic inflammation (Sect. 7.2), which is the central cause of many lifestyle-related diseases, such as insulin resistance (Sect. 9.2), T2D (Sect. 9.4) and atherosclerosis (Sect. 10.2) (Fig. 1.2). Moreover, also neurodegenerative disorders, such as Alzheimer's disease, most types of cancer, allergy, autoimmune diseases and inflammatory bowel diseases, such as Crohn's disease and ulcerative colitis, are closely linked to inflammation. Immune reactions in general, and inflammation in particular, are related to cellular metabolism. The proliferation of immune cells and their action in defense and tissue repair require high levels of energy metabolism. Thus, **metabolic stress, which is often caused by lipid overload in the blood and in adipose tissue (lipotoxicity), stimulates low-grade chronic inflammation**.

Among non-communicable diseases CVDs are the major contributor to the global burden of disease (46% worldwide, Sect. 10.1). Ischemic heart disease and stroke are the leading causes of death across all countries. Tobacco use, physical inactivity and unhealthy diet are responsible for about 80% of CHD (coronary heart disease) and cerebrovascular disease, *i.e.*, of heart attack and stroke. Ischemic disease or CHD, *i.e.*, the failure to supply oxygen to the heart muscle, is the major cause of CVD deaths (42.5%), while cerebrovascular disease, *i.e.,* the failure to supply oxygen to the brain, causes 35.5% of the CVD deaths. Atherosclerosis is the

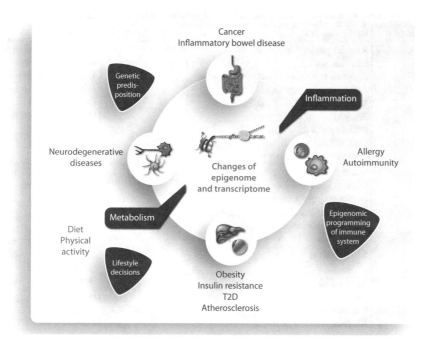

Fig. 1.2 Immune-mediated pathologies as key driver processes of diseases in various target organs. Inflammation and cellular metabolism are closely linked *via* coordinated changes in the epigenome and transcriptome of target tissues and cell types

basic pathophysiological lesion of CVDs, which tends to occlude the arteries to a varying extent (Sect. 10.2). A variety of cell types and lipids are involved in the pathogenesis of atherosclerotic plaques and arterial thrombosis.

There is convincing evidence that

- a plant-based diet composed of fruits, vegetables, leafy greens (polyphenols, anti-oxidants, folate, fibers, potassium etc.), nuts and seeds, such as walnuts, flaxseed and rapeseed oil (foods rich in the essential ω-3 fatty acid α-linolenic acid)
- fish and fish oil (containing the marine ω-3 fatty acids eicosapentaenoic acid and docosahexaenoic acid)
- physical activity, normal-range BMI and low alcohol intake
- avoiding SFAs (animal products, palm oil, coconut oil) and *trans*-fatty acids (hardened fats)

all contribute to the reduction of CVD risk (Table 1.3).

Table 1.3 Overview of lifestyle factors and risk of developing CVDs

Evidence	Decreased risk	No relationship	Increased risk
Convincing	Regular physical activity Linoleic acid Fish and fish oils (eicosapentaenoic and docosahexaenoic acid) Vegetables and fruits (including berries) Potassium Low to moderate alcohol intake (for coronary heart disease)	Vitamin E supplements	Myristic and palmitic acids Trans fatty acids High sodium intake Overweight High alcohol intake (for stroke)
Probable	α-Linolenic acid Oleic acid Non-starch polysaccharides Wholegrain cereals, nuts (unsalted) plant sterols/stanols, folate	Stearic acid	Dietary cholesterol Unfiltered boiled coffee
Possible	Flavonoids Soy products		Fats rich in lauric acid Impaired fetal nutrition Beta-carotene supplements
Insufficient	Calcium Magnesium Vitamin C		Carbohydrates Iron

1.5   Nutrition and Cancer

Cancer is the second leading cause of death globally. Unfortunately, every second of us will be diagnosed cancer at some point of his/her life. Cancer is defined as a disease, in which the normal control of cell division is lost, so that an individual cell multiplies inappropriately forming a primary tumor. The tumor cells may eventually spread through the body and form metastases. Cancer can arise from different tissues and organs, thus there are many different types of cancer. Some oncogenes and tumor suppressor genes encode for enzymes regulating the metabolism of nutrients, *i.e.*, the mutation of these genes can lead to the production of onco-metabolites affecting chromatin modifying enzymes (chromatin modifiers, Sect. 5.1). Thus, **dietary molecules affecting the epigenome can both increase or reduce the risk of cancer**. Migrant studies showed that moving from a region with low risk to one with a high risk leads within one generation to the same cancer pattern of the host country, *i.e.*, **the environment and lifestyle causes the cancer rather than the individual's genome**. Studies with monozygotic twins support these findings.

Approximately, **one third of cancer deaths are due to lifestyle choices, such as high BMI, low fruit and vegetable intake, lack of physical activity, tobacco and/ or alcohol use**. However, only a few definite relationships between specific nutrient-related factors and cancer are established. For example, there is convincing evidence that overweight and obese individuals have increased risk of cancer of esophagus, colorectum, breast (in post-menopausal women), pancreas, liver and

Table 1.4 Overview of lifestyle factors and risk of developing cancer

Evidence	Decreased risk	Increased risk
Convincing[a]	Physical activity (colon)	Overweight and obesity (esophagus, colorectum, pancreas, liver, breast in postmenopausal women, endometrium, kidney)
		Alcohol (oral cavity, pharynx, larynx, esophagus, liver, colorectum, breast postmenopausal)
		Processed meat (colorectum)
		Aflatoxin (liver)
Probable[a]	Fruits and vegetables (oral cavity, esophagus, stomach, colorectum**)	Red meat (colorectum)
	Physical activity (breast in postmenopausal women, endometrium)	Salt-preserved foods and salt (stomach)
	Whole grain, fiber, dairy products (colorectum)	Alcohol (breast in premenopausal women)
	Coffee (liver, endometrium)	Chinese-style salted fish (nasopharynx)
	Alcohol (kidney)	Glycemic load (endometrium)
Possible/ insufficient	Fiber	Animal fats
	Soya	Heterocyclic amines
	Fish	Polycyclic aromatic hydrocarbons
	ω-3 fatty acids	Nitrosamines
	Carotenoids	
	Vitamins B$_2$, B$_6$, folate, B$_{12}$, C, D, E	
	Calcium, zinc and selenium	
	Non-nutrient plant constituents (e.g., allium compounds, flavonoids, isoflavones, lignans)	

[a]The "convincing" and "probable" categories in this report is taken from the WCRF network report Recommendations and public health and policy implications 2018
[b]For colorectal cancer, a protective effect of fruit and vegetable intake has been suggested by many case-control studies but this has not been supported by results of several large prospective studies, suggesting that if a benefit does exist it is likely to be modest

kidney, while individuals who consume a high amount of alcohol are prone to cancers of the oral cavity, pharynx, larynx, esophagus, liver, colorectum and breast. Individuals who consume high amounts of processed meat have increased risk of colorectum cancer. Moreover, aflatoxins contribute to the development of liver cancer (Table 1.4). Importantly, there is convincing evidence that physical activity (Sect. 1.6) decreases the risk of colon cancer. Dietary factors that increase cancer risk include high intake of red meat (colorectum), salt-preserved foods (stomach), food with high glycemic index (endometrium) and alcohol (stomach, breast in premenopausal women). In contrast, protective factors are foods high in dietary fiber, such as whole grain products, fruits and vegetables (colorectal cancer), coffee (liver and endometrium cancers) and physical activity (endometrium and breast cancer in post-menopausal women).

Cancer and obesity are examples of non-communicable diseases, in which inflammation is part of the underlying cause of the disease (Sects. 7.4 and 8.3). WAT (White adipose tissue) is an important endocrine and metabolic organ consisting of both lipid-laden adipocytes and a stromal-vascular fraction, which contains preadipocytes, macrophages, other immune cells and endothelial cells (Fig. 1.3a). Obesity increases the size of adipocytes (hypertrophy) and number of adipocytes (hyperplasia) and is accompanied by infiltration of macrophages in the adipose tissue (Sect. 8.3). Elevated levels of circulating pro-inflammatory cytokines and acute phase proteins, such as CRP (C-reactive protein), characterize inflammatory responses that are triggered by obesity. In addition, increased release of pro-inflammatory adipokines (*i.e.*, hormones secreted by adipocytes), such as leptin, IL (interleukin) 6, resistin, SERPINE1 (serpin peptidase inhibitor, clade E, also called plasminogen activator inhibitor 1) and TNF (tumor necrosis factor), and reduced release of anti-inflammatory adipokines, such as adiponectin, is associated with obesity (Sect. 8.2).

The link between obesity and cancer initiation as well as the molecular mechanisms underlying how obesity converses normal epithelial cells to tumor cells is not completely understood (Fig. 1.3b). However, it is known that in cancer cells, in addition to low-grade inflammation and the release of inflammatory cytokines, also lipid metabolism is altered. Furthermore, obesity influences insulin signaling, which may provide further energy to cancer cells, *i.e.*, elevated insulin promotes their proliferation. For example, patients with insulin resistance have a poorer response to cancer treatment or bear a more aggressive cancer phenotype, probably due to their increased circulating insulin levels. Moreover, **elevated levels of leptin and reduced levels of adiponectin stimulate tumor growth**. Adiponectin signal transduction acts via transmembrane proteins activating the kinase AMPK (adenosine monophosphate-activated protein kinase, Sect. 6.6). AMPK is a critical negative regulator of proliferation in response to energy status as it induces growth arrest and apoptosis. Furthermore, adiponectin activates the nuclear receptor PPAR (peroxisome proliferator-activated receptor) α that controls fatty acid β-oxidation (Sect. 3.3).

1.6   Impact of Physical Activity

Some two million years ago, our ancestors made the unique adaption of "striding bipedalism", which significantly increased the capacity for long-distance walking and endurance running. These properties were essential for avoidance of predation, effective scavenging and persistent hunting. In parallel, humans lost most of their body hair, in order to allow better thermoregulation during this intense physical activity. Thus, **in the past efficient physical activity was essential for survival**. From the historical and evolutionary perspective, **an exercise-trained state is the biologically normal condition for humans**. However, in Western or Westernized societies a sedentary lifestyle is nowadays so widespread that often exercise is

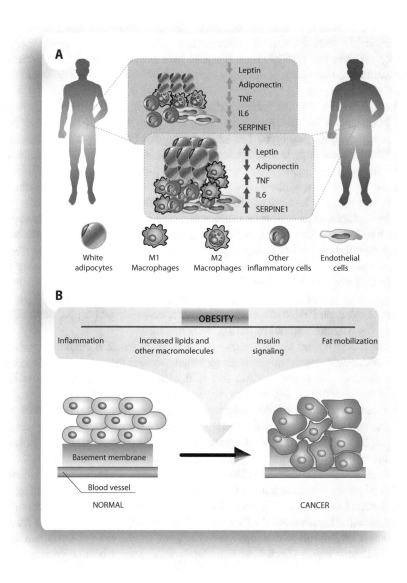

Fig. 1.3 Obesity, cancer and inflammation. WAT is an endocrine and metabolic organ consisting of lipid-laden adipocytes and preadipocytes, macrophages, other cells of the immune system and endothelial cells (**a**). In subjects with normal weight, the adipose tissue secretes high levels of adiponectin. During weight gain, WAT expands, which mediates the infiltration of macrophages and other inflammatory cells and leads to the secretion of the cytokine TNF from macrophages. Furthermore, the secretion of IL6, SERPINE1 and leptin is also increased. Elevated inflammation, increased availability of lipids and other macromolecules, impaired insulin signaling and changes in adipokine signaling leading to fat mobilization all contribute to the conversion of epithelial cells to an invasive tumor (**b**)

referred to as already having "health benefits". An inactive lifestyle and access to energy-dense nutrition over an extended lifespan increases the risk for non-communicable diseases, such as atherosclerosis, T2D and cancer. Thus, **physical activity reduces the risk of these diseases and prevents obesity**, since exercise increases the consumption of energy, *i.e.*, it burns off the body fat that would otherwise accumulate.

The increased metabolic activity of contracting skeletal muscles affects whole-body homeostasis by communicating with other organs, such as adipose tissue, liver, pancreas, bone and brain. For example, regular exercise promotes cardiovascular health, since it is beneficial for the profile of serum lipids *via* decreasing plasma triacylglycerols and increases of HDL (high-density lipoprotein) cholesterol (Sect. 10.3). In addition, physical activity also has an anti-inflammatory effect that can protect against low-grade chronic inflammation-associated diseases. Thus, **exercise-induced restoration or maintenance of metabolism and bioenergetics of the whole body changes homeostatic signals affecting nutrient uptake and growth factor availability in a multitude of tissues, both in health and disease.** Reliable epigenetic biomarkers for the assessment of the interindividual variation of the effect of exercise training in the prevention and therapy of metabolic diseases are desired. A promising candidate is the epigenetic modification of regulatory regions of the gene *PPARGC1A* (PPAR gamma, co-activator 1 alpha, Sect. 6.2) *via* DNA methylation in skeletal muscle of T2D patients.

Physical activity has also a positive effect on disorders that are not directly connected with energy metabolism, such as cancer, mental disorders and neurodegenerative diseases. To a large extent this is due to the anti-inflammatory effect of exercise that causes a higher production and release of myokines (*i.e.*, anti-inflammatory cytokines of skeletal muscle) and the reduced expression of TLRs (Toll-like receptors) in associated immune cells. TLRs are PRRs (pattern recognition receptors) on the surface of monocytes and macrophages that detect pathogens and initiate the innate immune response (Sect. 7.2). These anti-inflammatory effects inhibit the production of pro-inflammatory cytokines. Exercise also reduces the number of pro-inflammatory monocytes and increases the count of circulating regulatory T cells (T_{REG}). T_{REG} are a specialized subpopulation of T cells suppressing the activation of the immune system (Sect. 7.1). Since regular exercise reduces fat mass, there is less infiltration of macrophages to adipose tissue and a switch from M1- to M2-type macrophages (Sect. 7.4). This leads to an increase in adiponectin levels and a decrease in pro-inflammatory adipokines, such as IL6, TNF and leptin, in the circulation. Physical exercise also influences the CNS. Impulses from the brain and contracting muscles elevate plasma cortisol and adrenaline production in adrenal glands. These hormones suppress inflammation by decreasing the production of pro-inflammatory cytokines by monocytes and macrophages.

The worldwide obesity epidemic (Sect. 8.1) **is paired with too low physical activity among the concerned persons**. Therefore, pharmacologic intervention with exercise mimetics is considered. These molecules can mimic the effect of exercise by acting as key components of exercise-induced muscle adaptation, such as mitochondrial remodeling and bioenergetics. In this way, they are able to produce the benefits of fitness, such as increased mitochondrial oxidative phosphorylation and fatty acid metabolism leading to lower blood glucose levels, reduced inflammation and increased endurance. For example, the synthetic compounds AICAR and GW501516 activate the kinase AMPK and the nuclear receptor PPARδ, respectively. Both proteins play key roles in the pathways of mitochondrial biogenesis and fatty acid β-oxidation. This induces energy expenditure without any change in activity, *i.e.*, it allows calorie burning without physical activity. Mechanistically, this works via the induction of the UCP (uncoupling protein) 2 and UCP3, which convert the H^+ gradient at the inner mitochondrial membrane into heat (Sect. 8.2). Although the potential of exercise mimetics to efficiently promote fat burning represents an interesting medical opportunity, **there is also the risk that the compounds are used for doping of endurance athletes**.

Additional Readings

Afshin A, Sur PJ, Fay KA, Cornaby L, Ferrara G, Salama JS, Mullany EC, Abate KH, Abbafati C, Abebe Z et al (2019) Health effects of dietary risks in 195 countries, 1990–2017: a systematic analysis for the global burden of disease study 2017. Lancet 393:1958–1972

Blüher M (2019) Obesity: global epidemiology and pathogenesis. Nat Rev. Endocrinol 15:288–298

Fan W, Evans RM (2017) Exercise mimetics: impact on health and performance. Cell Metab 25:242–247

Hawley JA, Hargreaves M, Joyner MJ, Zierath JR (2014) Integrative biology of exercise. Cell 159:738–749

Koelwyn GJ, Quail DF, Zhang X, White RM, Jones LW (2017) Exercise-dependent regulation of the tumour microenvironment. Nat Rev. Cancer 17:620–632

NCD-Risk-Factor-Collaboration (2017) Worldwide trends in body-mass index, underweight, overweight, and obesity from 1975 to 2016: a pooled analysis of 2416 population-based measurement studies in 128.9 million children, adolescents, and adults. Lancet 390:2627–2642

Chapter 2
Human Genomic Variation

Abstract This chapter will briefly describe the genetic adaption of anatomically modern humans due to migration to new geographic and climatic environments in Asia and Europe. This includes also the challenges provided by the shift from hunters and gatherers to farmers. **Genetic differences between human populations are most pronounced in tissues, such as the skin, the intestinal tract or the immune system, that are directly affected by the environment**. This led not only to obvious differences in skin color among the populations, but also in different resistance to diseases and diversity in dietary intake, such as the ability to digest lactose (milk sugar). The genetic basis of the variation of human populations and individuals has recently been studied and catalogued by large consortia, such as the *1000 Genomes Project*. Genome-wide genotyping and whole genome sequencing allows the study and analysis of complex diseases, such as T2D and CVDs, on the basis of dozens to hundreds of genetic variants, such as SNPs and CNVs (copy number variations).

Keywords Human evolution · Human populations · Single nucleotide variants · Copy number variants · Haplotype blocks · Next-generation sequencing · *HapMap Project* · Genome-wide association studies · *1000 Genomes Project*

2.1 Migration and Evolutionary Challenges of *Homo sapiens*

Approximately 300,000 years ago anatomically modern humans developed in East Africa. **The main characteristic of *Homo sapiens* is a superior locomotive ability that is essential for encountering predators and food procurement**. Some 50–75,000 years ago, reasonable numbers of these modern humans started to migrate to Asia and Europe and replaced there through interbreeding already prevalent archaic (*i.e.*, nowadays extinct) human species, such as the Neanderthals (Fig. 2.1). Due to their new environments our ancestors were exposed to a number of divergent selective pressures, such as thermoregulation in colder climates, tolerance to hypoxia (*i.e.*, oxygen supply deprivation) at high altitude and light skin pigmentation in regions with lower levels of UV-B (ultraviolet B radiation, Sect. 4.1).

© Springer Nature Switzerland AG 2020
C. Carlberg et al., *Nutrigenomics: How Science Works*,
https://doi.org/10.1007/978-3-030-36948-4_2

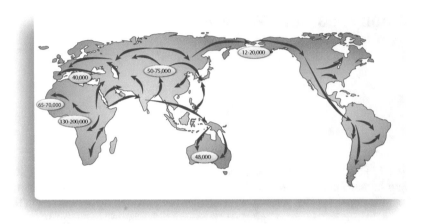

Fig. 2.1 Migrations of *Homo sapiens*. The spread of anatomically modern humans from East Africa over the rest of the continent was followed by an expansion from the same area to Asia, probably by both a southern and northern route some 50–75,000 years ago. Oceania, Europe and the Americas were settled from Asia in that order. The migration patterns are primarily based on analyses of changes in mitochondrial DNA

Some 10,000 years ago our ancestors started to give up their hunter and gatherer habit and became farmers. This significant lifestyle change was associated with distinct foods, such as cereals and milk (Sect. 1.1). **The improved nutrition supply allowed higher population densities but was compromised with an increased load with infectious diseases**, many of which were acquired from domesticated animals. Both dietary changes and immunological challenges caused dominant evolutionary pressure and rather rapid genetic adaption. The phenotypic consequences of these genetic adaptations did not only shape the biological variation but also the health and disease risk of the 7.8 billion people presently living on Earth.

In chap. 1 we already started to discuss the significant increase in obesity and the resulting increase in the rates of cancer and metabolic diseases. Worldwide it took several thousand years, *i.e.*, clearly more than 100 generations, to turn most humans from hunters and gatherers to farmers but only less than 50 years to be preferential users of cars, supermarkets and fast-food. This means that **humans simply had no time to adapt genetically to the rapidly changing "obesogenic environment"**. In the context of an inactive lifestyle combined with energy-dense foods, the genetic variations that had been initially evolved for an efficient energy storage and physical mobility turned to be an increased risk for developing chronic non-communicable diseases, such as T2D or CVDs (Chap. 9 and 10).

2.2 Diversity of Human Populations

When humans became distributed to the different continents, i.e. geographically isolated, new gene variations could not be spread anymore to all members of the species. Since in the past distant human populations were less likely to exchange migrants, they cluster genetically in relation to their geographic distance from each other. Thus, human genetic variation diverted geographically, when individuals accumulated further mutations during the past 50,000 years. Since anatomically modern humans lived in Africa already for some 300,000 years, populations on this continent are more diverse than in the rest of the world. **Although there are obvious phenotypic differences of populations concerning skin color, body height and facial features, there are no absolute genetic differences between them**. For example, there is no single nucleotide difference that can distinguish Africans from Eurasians, but population differences are based on thousands of gene loci. This implies that a property (often referred to as a "trait"), such as skin color (Sect. 4.1), can change rather rapidly, when the allele frequencies shift at the respective loci contributing to the trait (Box 2.1). Some 500 years ago, when navigation over the oceans became possible, voluntary and involuntary (slaves) migration started, which caused significant population admixtures, particularly in the Americas, but also in other continents. Before that time there had been at least two major events of admixture in Europe, where first some 9000 years ago early farmers from Anatolia interfered with indigenous European hunters and gatherers and then some 5000 years ago Yamnaya pastoralists from the Eurasian steppe migrated to Europe, *i.e.*, the phenotype of present-day Europeans is largely the product of this Bronze Age collision of this three ancestral tribes.

Box 2.1 Natural selection
Positive natural selection, *i.e.*, the force that drives the increase in prevalence of advantageous traits, has played a central role in human evolution. Individuals with advantages (referred to as "adaptive traits") tend to be more successful in reproduction, *i.e.*, they contribute with more offspring to the following generation than others. Due to inheritance from one generation to the other, the selection process increases the prevalence of the respective adaptive trait. Under persistent selection pressure such adaptive traits, step by step, may become universal to the population. **Factors fostering selection, *i.e.*, evolutionary pressures include, *e.g.*, limits on resources, such as food, or threats, such as pathogens**. However, adaptive traits not always become prevalent within a population. Gene frequency alteration in a population can also happen *via* a genetic drift of genes that are not under selection (Sect. 5.4). In this context, even a deleterious allele may become universal to the members of a population, *e.g.*, under the influence of a weak selection or in small populations.

Hundreds of complex phenotypic traits determine how an individual looks and behaves as well as his/her risks to develop non-communicable diseases. Furthermore, each complex trait is based on dozens to hundreds of gene variants and environmental influences, *i.e.*, for an understanding the molecular basis of a trait its genetic architecture needs to be uncovered, such as

- the number of variants that influence a heritable phenotype
- their relative magnitude concerning different traits
- the population frequency of the respective variants
- their interactions with each other and the environment.

Whole genome sequencing has indicated that **every individual carries, on average, 4–5 million genetic variants** (Sect. 2.4) covering about 12 Mb (million base pairs) of sequence (0.3% of all). Most of these genetic variants are neutral, *i.e.*, they do not contribute to phenotypic differences or disease risk, and are achieved simply by chance significant frequencies within respective human populations. In this context, the sum of rare, high-penetrance variants is of significant influence. However, there is a large number of common variants with a small to modest effect size that have a dominant role in common complex traits. For example, the trait "body height" is dependent on at least 180 gene loci, *i.e.*, it is a prime example of a complex/polygenic trait (for further examples see Chaps. 8, 9 and 10). In Europe, this trait has changed significantly (in average by 10 additional centimeters) within the last few generations under the environmental trigger of improved quality and quantity of nutrition.

During evolution of *Homo sapiens* up to 10% of all protein-coding genes, *i.e.*, some 2000 genes, may have been affected by positive natural selection. In particular, the immune system, the digestive tract and the skin (including hair, sweat glands and sensory organs) had been susceptible to positive selection (Sect. 4.1). This is due to the fact that these organ systems are more likely in contact with the environment than other parts of our body. For example, variants of the innate and adaptive immune system (Box 2.2), such as genes encoding for membrane immune receptors, had been under special positive selection by pathogenic microbes. An interesting example is CCR5 (C-C chemokine receptor 5), which is essential for the entry of HIV (human immunodeficiency virus) 1 into T cells, since a 32 bp deletion in the *CCR5* gene protects its carriers from HIV infection. This mutation is currently becoming positively selected in populations where HIV1 infections occur on larger scale, such as in South Africa. Moreover, some alleles that were introduced into modern humans through interbreeding with archaic human species have been positively selected. For example, a Neanderthal variant of the *SLC16A11* (solute carrier family 16 member 11) gene, which encodes for a lipid transporter in the ER (endoplasmic reticulum), reached high frequencies, *e.g.*, in native Americans, where it is associated with increased T2D risk.

Box 2.2 The innate and adaptive immune system
In general, the immune system is composed of biological structures, such as lymph nodes, cell types, such as monocytes, macrophages, T and B lymphocytes (*i.e.*, cellular immunity), and proteins, such as complement proteins and antibodies (*i.e.*, humoral immunity), that protect the organism against infectious diseases. The immune system detects a wide variety of molecules, known as antigens, of potential pathogenic origin, such as on the surface of microbes, and distinguishes them from the organism's own healthy tissue. The immune system is classified into innate and adaptive immunity. The innate immune system is evolutionary older, bases on monocytes, macrophages, neutrophils and natural killer cells, and uses destructive mechanisms against pathogens, such as phagocytosis, with the support of anti-microbial peptides from the complement system. Adaptive immunity applies more sophisticated defense mechanisms, in which T and B cells use highly antigen-specific surface receptors, such as T cell receptors and B cell receptors, the latter finally turning into secreted antibodies. Moreover, after an initial specific response to a pathogen the adaptive immune system creates an immunological memory that leads to an enhanced response to subsequent encounters with that same antigen.

2.3   Genetic Variants of the Human Genome

The reference haploid sequence of the human genome (Box 2.3) was released in 2001 by the first "big biology" project, the *Human Genome Project* (Box 2.4), and reflects the assembly of sequences derived from a few male donors. SNPs are variants of the sequence of the reference genome where exactly one nucleotide (A, T, G or C) is altered (Fig. 2.2). In contrast, structural variants of the genome mostly affect more than one nucleotide. These can be insertion-deletion (indel) variants, where in most cases only a few bases are added or removed, respectively, but there are also indels of up to 80 kb (kilo bp) in length. Indels that are not multiples of 3 bp in length and are located within protein coding regions result in frameshift mutations, *i.e.*, from the position of the mutation onwards the whole amino acid sequence of the encoded protein is changed. Furthermore, CNVs consist of deletions or insertions of DNA stretches in one genome compared to another. These variants can be heterozygous or homozygous. A predominant class of insertions is that of ancient transposons. These DNA stretches persist in the genome as SINEs (short interspersed elements, *e.g.*, *Alu* elements) and LINEs (long interspersed nuclear elements).

Box 2.3 The human genome
The human genome is the complete sequence of the anatomically modern human and was obtained by the *Human Genome Project* (www.genome. gov/10001772) *via* whole genome sequencing. This reference sequence is accessible *via* different genome browsers (*e.g.*, genome.ucsc.edu or www. ensembl.org) and represents the assembly of the genomes of a few young healthy male donors. With the exception of germ cells, *i.e.,* female oocytes and male sperm, each human cell contains a diploid genome formed by 2x 3.235 billion bp (3235 Mb) that is distributed on 2x 22 autosomal chromosomes and two X chromosomes for females and a XY chromosome set for males. In addition, every mitochondrion contains 16.6 kb mitochondrial DNA. The haploid human genome contains some 20,000 protein-coding genes and about the same number of non-coding RNA (ncRNA) genes. The protein-coding sequence covers less than 2% of our genome, *i.e.*, **the 98% of the genome is non-coding and primarily has regulatory function**.

Almost 50% of the sequence of our genome is formed by repetitive DNA (often also referred to as "junk DNA"), which is sorted into the following categories (by order of frequency):

LINEs (500–8000 bp)	21%
SINEs (100–300 bp)	11%
Retrotransposons, such as LTRs (long terminal repeats) (200–5000 bp)	8%
DNA transposons (200–2000 bp)	3%
Minisatellite, microsatellite or major satellite (2–100 bp)	3%

LINEs and SINEs are identical or nearly identical DNA sequences that are separated by large numbers of nucleotides, *i.e.*, the repeats are spread throughout the whole human genome. LTRs are characterized by sequences that are found at each end of retrotransposons. DNA transposons are full-length autonomous elements that encode for a transposase, *i.e.*, an enzyme that transposes DNA from one to another position in the genome (also known as "jumping DNA").

Box 2.4 Big biology projects

With a delay of some 20 years molecular biologists followed the example of physicists and realized that some of their research aims could only be reached through multinational collaborations of dozens to hundreds of research teams and institutions in so-called big biology projects. The *Human Genome Project* (www.genome.gov/10001772), which was launched in 1990 and finished in 2003, was the first example. Together with follow-up studies the project had a tremendous impact on the understanding of the architecture and function of human genes. The *HapMap Project* (http://hapmap.ncbi.nlm.nih.gov) was one of these follow-ups benefitting from advancing genotyping technologies. In parallel, improved next-generation sequencing methods allowed personal genome sequencing of both normal and cancer genomes. This made large-scale genome sequencing studies possible, such as the *1000 Genomes Project* (www.1000genomes.org) and *The Cancer Genome Atlas* (TCGA) (www.cancer.gov/about-nci/organization/ccg/research/structural-genomics/tcga). Furthermore, the *Encyclopedia of DNA elements (ENCODE) Project* (www.genome.gov/encode) and the *Functional Annotation of the Mammalian Genome (FANTOM5) Project* (http://fantom.gsc.riken.jp/5) focused on the functional characterization of the human genome. The *ENCODE* follow-up project *Roadmap Epigenomics* (www.roadmapepigenomics.org) provided human epigenome references from 111 primary human tissues and cell lines.

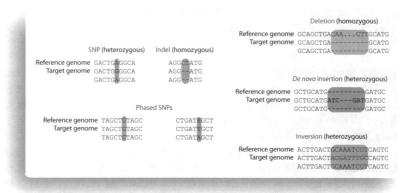

Fig. 2.2 Types of variations present in human genome sequences. The haploid reference genome is indicated at the top of each variant example, while the individual's diploid genome is shown below. The genetic variants can be either heterozygous or homozygous. Phased SNPs refers to their order within a haplotype

The different types of human genetic variants are referred to as common (or polymorphisms), when they have a MAF (minor allele frequency) of at least 1% within the studied population, or as rare, when they have a MAF of less than 1%. SNPs represent the most common class of genetic variations among individuals and approximately seven million SNPs show a MAF of more than 5% (www.ncbi.nlm.nih.gov/SNP). The *1000 Genomes Project* (Sect. 2.5) indicated that in addition there is a huge number of rare and novel SNPs. Nevertheless, the majority of variants of any given individual are common in the whole population. In parallel, some 60,000 unique CNVs are known and some of them are quite common in human populations. Since the detection of structural variants needs advanced technology, basically all initial associations between genome variations and complex traits, such as observed by GWAS (genome-wide association studies), were done only with SNPs. Nevertheless, per individual structural variants cover between 9 and 25 Mb of sequence, *i.e.*, 0.3–0.8% of the whole genome.

The average difference in nucleotide sequence of a pair of familial unrelated humans lies in the order of 1 in 1000. This proportion is low compared with other species and **confirms the recent origin of *Homo sapiens* from a small founding population**. The impact of SNPs on the coding sequence of the human genome is well established. Synonymous mutations do not alter the encoded protein, while non-synonymous mutations cause a change in the amino acid sequence (missense) or introduce a premature stop codon (nonsense). Indels as well as CNVs in exonic sequences can result in either non-frameshift or frameshift mutations. Moreover, CNVs in intronic sequences may lead to alternative splicing. The impact of genetic variations in the non-coding region of the human genome will be discussed in Sect. 4.3.

2.4   Haplotype Blocks and GWAS

The genetic approach of linkage analysis was used in the past, in order to identify genes responsible for monogenetic disorders, such as the neurodegenerative Huntington's disease. However, these rare diseases represent only a relatively small fraction of all disorders. In contrast, **most human diseases have a complex origin**, *i.e.*, they involve many gene loci in a complex interaction pattern (Sect. 4.2). For these cases the genome-wide identification of SNPs *via* GWAS was found to be a more suitable approach. With an average SNP density of 1 in 1000 nucleotides these studies require the testing of millions of SNPs per individual and hundreds to thousands of subjects. For this purpose, large-scale studies (Box 2.4), such as the *Human Genome Diversity Project* and the *HapMap Project*, had been launched. They used high-throughput SNP genotyping technologies, such as arrays with up to a million SNPs. *HapMap* started in 2002 with 270 samples from the three major human populations, which were

- 90 samples from Yoruba individuals living in Ibadan, Nigeria

- each 45 samples from Han Chinese individuals living in Beijing, China, and Japanese individuals living in Tokyo, Japan
- 90 samples from individuals with European ancestry living in Utah, USA.

In its latest version, *HapMap 3*, the study extended to 1184 individuals from 11 global populations. The consortium performed genotyping for 1.6 million common SNPs and CNVs and used knowledge from linkage disequilibrium analysis of haplotype blocks.

Haplotype blocks are stretches of genomic DNA of typically 10–100 kb in length that are inherited from generation to generation in blocks, *i.e.*, they are not interrupted by meiotic recombination events (Fig. 2.3). In contrast, the borders of haplotype blocks represent recombination events that had happened many generations ago in our ancestors. In fact, **gene conversion during meiosis is some 100 times more frequent than point mutations and therefore a more efficient mechanism of evolution**. For this reason, sexual reproduction evolved. Since African populations have existed far longer than European and Asian populations, their haplotype blocks are shorter, *i.e.*, the blocks had more time to decay because of the accumulation of recombination events in a higher number of generations. In contrast, all non-African humans derived from a small population of eastern African origin, *i.e.*, they went through a demographical bottleneck that is clearly visible in the limited diversity of their genomes.

GWAS employs an "agnostic" approach in the search for unknown disease variants, *i.e.*, they interrogate a large number of SNPs covering the entire human genome. 15 years of intensive GWAS research resulted in more than 4000 publications reporting some 150,000 SNPs (as of August 2019) being statistically robustly associated with one out of more than 500 complex diseases and traits (*Catalog of Published Genome-Wide Association Studies*, www.ebi.ac.uk/gwas). Nowadays, GWAS meta-analyses include more than 100,000 individuals and can explain a larger proportion of trait heritability. For example, the heritability of about 20% of CHD cases is explained by some 80 genetic loci, 20% of T2D by some 100 loci, 20% of inherited breast cancer by some 150 loci, 33% of inherited prostate cancer by some 100 loci and 30% of Alzheimer's disease by some 20 loci.

Despite this notable success, in average **not much more than 10% of the heritability of most complex, polygenic traits have been explained by common variants** assessed by GWAS. In general, the missing or unsolved heritability does not allow assigning an individual with any reliable estimation about his/her risk for a particular disease, when exclusively SNP analysis is performed (for more advanced analyses, see Sect. 4.5). The only well-known exceptions are age-related macular degeneration and type 1 diabetes (T1D), for which the combinations of common and rare variants can provide a quantifiable risk profile. GWASs with 2000–5000 individuals confidently identify common variants with effect sizes, referred to as ORs (odds ratios), of 1.5 or greater, *i.e.*, the odds of having, *e.g.*, T1D are 1.5 times higher the odds of being without the disease when carrying common variants. This makes it unlikely that further common SNPs with moderate or even large ORs in complex traits will be discovered in future. Limited statistical power to detect small

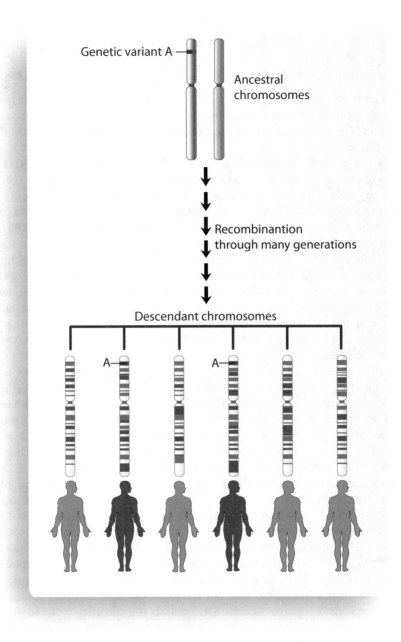

Fig. 2.3 The origin of haplotypes. Two plain ancestral example chromosomes get scrambled through meiotic recombination over many generations, in order to yield different descendant chromosomes. For example, after 30,000 years a typical chromosome will have undergone more than one crossover per 100 kb. In the case of a genetic variant (marked by an A) on one ancestral chromosome the risk of a particular disease increases. Thus, the two individuals (red) in the current generation who inherited that region of the ancestral chromosome will be at increased risk. Within the haplotype block that carries the disease-causing variant there are many SNPs that can be used to identify the location of the variant

gene-gene and gene-environment interactions requires increasing sample sizes, which may be achieved by pooling several GWASs through meta-analysis. For example, sample sizes of 60,000 subjects are necessary to provide sufficient power to identify the majority of variants with ORs of 1.1 (*i.e.*, a 10% increased risk for the tested trait). Some of the missing heritability may be explained by rare variants with high ORs, which are poorly captured by standard GWASs. In addition, **environmental exposures, including those experienced as fetus** (Sect. 5.3)**, affect the epigenome and may explain large parts of the missing heritability**.

For any investigated trait or disease a larger number of SNPs with high linkage disequilibrium to the genomic variants are described on the same haplotype block, but only a very few of them may explain the mechanistic basis of the trait. This implies that most disease-associated SNPs are not functionally relevant to mechanisms of the respective trait. Moreover, extensive sequencing of associated regions may identify additional, previously unknown, rare variants with a possible biologic role. This was one of the goals of the *1000 Genomes Project* (Sect. 2.5). However, only for protein-coding regions, which carry some 12% of all trait-associated SNPs, a straightforward mechanistic explanation of the impact of the genetic variation is possible, *i.e.*, many of the remaining 88% variations are regulatory SNPs affecting the genomic binding sites of transcription factors (Sect. 4.3).

2.5   The *1000 Genomes Project*

Important technological advances in high-throughput sequencing have led to a rapid decrease in the costs of DNA sequencing. As a result, whole genome sequencing became an affordable tool in understanding the genomic basis of health and disease. At present, already huge amounts of data have been obtained from whole genomes of both healthy and diseased individuals. This information has not only helped in disease stratification and in the identification of their molecular mechanisms, but also is **transforming the perspective of future health care from disease diagnosis and treatment to personalized health monitoring and preventive medicine** (Sect. 4.5).

Whole genome sequencing results in the identification of the complete set of genetic variants of a given individual (Fig. 2.4). As a natural extension of the *HapMap Project*, the *1000 Genomes Project* was initiated in 2008 with the aim to sequence the genomes of at least 1000 individuals from different populations around the world. *HapMap* populations were included in the project (Table 2.1), and in total 2504 genomes from 26 populations covering all five continents were investigated. In total, the project describes 88 million genome variants, of which 84.7 million are SNPs, 3.6 million short indels and 60,000 structural variants. Individuals from African ancestry populations show most variant sites confirming the out-of-Africa model of human origin (Sect. 2.1). A typical human genome carries 200,000 variants, most of which are common and only 20% are rare (MAF < 0.5%). In average, a typical genome contains some 150 variants resulting in protein truncation, 10,000

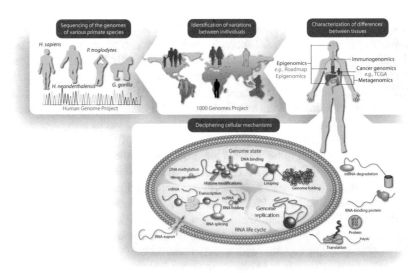

Fig. 2.4 Roadmap of sequencing science. The *Human Genome Project* (Box 2.4) created a reference genome. Nowadays, also the genomes of all other primate species are known including some extinct human species (**top left**). Whole genome sequencing of several thousand individuals was performed in large consortia, such as the *1000 Genomes Project* (**top center**). Moreover, the genetic and epigenetic differences between tissues and cell types of the same individual were collected in cancer genomics and epigenomics projects, such as *TCGA* and the *Roadmap Epigenomics* (**top right**). The application of different next-generation sequencing methods allows integrating many different processes within the cell (**bottom**)

changing amino acids and 500,000 affecting transcription factor binding sites. Interestingly, **each individual is heterozygous for 50–100 genetic variants that can cause inherited disorders in homozygous offspring**. This will provide a large demand and challenge for genetic counseling based on whole-genome sequencing. Moreover, gene-environment interactions provided by lifestyle factors, such as the personal choice of food, will create an additional level of complexity (Sect. 4.5).

The rapid maturation of next-generation sequencing technologies led to the exponential development of methods for nearly all aspects of cellular processes, *i.e.*, sequencing allows their detailed and comprehensive analysis. For example, epigenome-wide methods investigate various aspects of chromatin biology, such as DNA methylation, histone modification state and 3-dimensional (3D) chromatin structure (Fig. 2.4, Sect. 5.1). Next-generation sequencing methods have the advantage that they provide, in an unbiased and comprehensive fashion, information on the entire epigenome. Individual research teams as well as large consortia, such as the *ENCODE Project* and the *Roadmap Epigenomics Project* (Box 2.4), have already produced thousands of epigenome maps from hundreds of human tissues and cell types. The integration of these data, *e.g.*, transcription factor binding and characteristic histone modifications, allows the prediction of enhancer and promoter

Table 2.1 Populations of the *1000 Genomes Project*. Human populations that were included in the *1000 Genomes Project* are listed. The numbers of individuals that were investigated deep-coverage sequencing (*i.e.*, some 50 unique reads) of 2504 subjects are indicated. Individuals across 26 populations were sequenced. These populations were classified into 5 major continental groups (red): East Asia (EAS), Europe (EUR), Africa (AFR), America (AMR) and South Asia (SAS). Original *HapMap* populations are highlighted in bold

Full population name		Code	Individuals
Han Chinese in Beijing, China		**CHB**	**103**
South Han Chinese		CHS	105
Chinese Dai in Xishuangbanna, China	EAS	CDX	93
Japanese in Tokyo, Japan		**JPT**	**104**
Kinh in Ho Chi Minh City, Vietnam		KHV	99
Utah residents (CEPH) with Northern and Western European ancestry		**CEU**	**99**
Toscani in Italy		TSI	107
British in England and Scotland	EUR	GBR	91
Finnish in Finland		FIN	99
Iberian Populations in Spain		IBS	107
Yoruba in Ibadan, Nigeria		**YRI**	**108**
Luhya in Webuye, Kenya	AFR	LWK	99
Gambian in Western Division, Mandinka		GWD	113
Mende in Sierra Leone		MSL	85
African Ancestry in Southwest US		ASW	61
Esan in Nigeria		ESN	99
African Caribbean in Barbados		ACB	96
Mexican Ancestry in Los Angeles, CA, USA	AMR	MXL	64
Colombian in Medellin, Colombia		CLM	94
Peruvian in Lima, Peru		PEL	85
Puerto Rican in Puerto Rico		PUR	104
Bengal in Bangladesh		BRB	86
Gujarati Indians in Houston, TX, USA		GIH	103
Indian Telugu in the UK	SAS	ITU	102
Sri Lankan Tamil in the UK		STU	102
Punjabi in Lahore, Pakistan		PJL	96
		TOTALS	**2504**

regions as well as monitoring their activity and many additional functional aspects of the epigenome. In Chap. 4 we will discuss further applications of these projects including the identification and characterization of regulatory SNPs (Sect. 4.3) and the iPOP (integrative personal omics profile) of individuals (Sect. 4.5).

Additional Readings

Genomes Project C, Auton A, Brooks LD, Durbin RM, Garrison EP, Kang HM, Korbel JO, Marchini JL, McCarthy S, McVean GA et al (2015) A global reference for human genetic variation. Nature 526:68–74

Pääbo S (2014) The human condition-a molecular approach. Cell 157:216–226

Reich D (2018) Who we are and how we got here: ancient DNA and the new science of the human past. Oxford University Press, Oxford ISBN 978-0-19-882125-0

Tam V, Patel N, Turcotte M, Bosse Y, Pare G, Meyre D (2019) Benefits and limitations of genome-wide association studies. Nat Rev Genet 20:467–484

Timpson NJ, Greenwood CMT, Soranzo N, Lawson DJ, Richards JB (2018) Genetic architecture: the shape of the genetic contribution to human traits and disease. Nat Rev Genet 19:110–124

Veeramah KR, Hammer MF (2014) The impact of whole-genome sequencing on the reconstruction of human population history. Nat Rev Genet 15:149–162

Chapter 3
Sensing Nutrition

Abstract This chapter will describe distinct mechanisms of sensing the abundance of fatty acids, amino acids and glucose *via* membrane receptors, metabolic enzymes, regulatory kinases and transcription factors. The latter, in particular members of the nuclear receptor superfamily, play a key role in nutrient-sensing pathways. Many nuclear receptors bind macro- and micronutrients or their metabolites, such as fatty acids to PPARs, oxysterols to LXRs (liver X receptors) and vitamin D metabolites to VDR (vitamin D receptor), *i.e.*, **nuclear receptors are able to translate nutrient fluctuations into responses of the genome**. In metabolic organs nuclear receptors respond to nutrient changes and specifically activate hundreds of their target genes. Moreover, also the immune system is triggered in its inflammatory and antigen response by nuclear receptors and their ligands. In addition, nuclear receptors belong to those transcription factors that play a central role in managing the circadian clock both in the CNS as well as in peripheral organs. Basically all tissues and cell types of our body display a functional molecular clock, the coordination of which is essential for optimal physiology including metabolism.

Keywords Lipid sensing · Amino acid sensing · Glucose sensing · Nuclear receptors · Gene regulation · Lipid metabolism · PPARs · LXRs · VDR · Innate immunity · Adaptive immunity · Circadian clock

3.1   Nutrient-Sensing Mechanisms

Periodical scarcity of nutrients was a strong evolutionary pressure to select efficient mechanisms of nutrient sensing. This sensing process may be either the direct binding of the macro- or micronutrient to its sensor or an indirect mechanism that is based on the detection of metabolite that reflects the nutrient's abundance. The respective sensor is a protein that binds the nutrient with an affinity in the order

© Springer Nature Switzerland AG 2020
C. Carlberg et al., *Nutrigenomics: How Science Works*,
https://doi.org/10.1007/978-3-030-36948-4_3

of the fluctuations of its physiological concentrations. The sensing of nutrients may then trigger the release of hormones or other signaling molecules to the circulation, in order to achieve a coordinated response of the whole organism.

Due to their non-polar nature, *i.e.*, their insolubility in aqueous solutions, lipids are rarely found free in soluble form. For example, in serum they are either transported in lipoproteins and chylomicrons (Sect. 10.3) or bound by albumin. GPRs (G protein-coupled receptors), such as GPR40 and GPR120, bind long-chain unsaturated fatty acids, *e.g.*, in the membrane of β cells of the pancreas, and enhance in these cells glucose-triggered insulin release (Fig. 3.1a). In enteroendocrine cells of the intestine the binding of lipids to GPRs leads to the release of incretins (*i.e.*, gastrointestinal hormones that amplify insulin secretion). In the intestinal lumen, fatty acids are bound by the scavenger receptor CD36 (CD36 molecule) that initiates their uptake (Fig. 3.1b). In taste buds CD36 triggers calcium release from the ER and neurotransmission, while in enterocytes it promotes fatty acid uptake.

Adequate sensing of internal cholesterol levels is important, in order to avoid, in case of abundant external supply, the activation of the energetically demanding cholesterol biosynthetic pathway and to prevent toxic levels of free cholesterol in the cell. Cholesterol binds to the protein SCAP (SREBF chaperone) (Fig. 3.1c). In case of high intracellular cholesterol levels, SCAP increases its affinity for INSIG1 (insulin-induced gene 1) protein that anchors SCAP and the transcription factor SREBF1 (sterol regulatory element-binding transcription factor 1) within the ER membrane. When cholesterol levels are low, the SCAP-SREBF complex dissociates from INSIG and shuttles to the Golgi apparatus, where SREBF is released, translocates to the nucleus and activates genes involved in lipid anabolism, *i.e.*, cholesterol biosynthesis and lipogenesis (Fig. 3.1d). In addition, at low cholesterol levels the enzyme HMGCR (HMG-CoA reductase), which is also located in the ER membrane, catalyzes a rate-limiting step of the cholesterol *de novo* synthesis (Fig. 3.1e). In contrast, high levels of intermediates in the cholesterol biosynthesis pathway, such as lanosterol, trigger the binding of HMGCR to INSIG, which leads to the ubiquitin-mediated degradation of the enzyme (Fig. 3.1f). LXRs are activated by elevated cholesterol levels (Sect. 3.4), *i.e.*, the pathways of SREBF and LXRs work in a reciprocal fashion, in order to maintain cellular and systemic cholesterol homeostasis.

In case of shortage in amino acids, cellular proteins are used as a reservoir and degraded *via* the proteasome or by autophagy (Box 3.1). The latter is a basic catabolic mechanism that involves the degradation of unnecessary or dysfunctional cellular components through the actions of lysosomes. The lysosome is an organelle, in which amino acids and other nutrients are scavenged from cellular components. Accordingly, a protein located at the outer surface of lysosomes, the TOR (target of rapamycin) complex 1 (TORC1), acts as the key amino acid sensor (Sect. 6.1). Thus, **high levels of amino acids within the lysosome reflect their abundance in the cell**. This process promotes survival during starvation by maintaining cellular energy levels. In contrast, during periods of prolonged starvation and hypoglycemia, amino acids can be catabolized for the production of glucose and ketone bodies, *i.e.*, they provide essential energy sources for the brain.

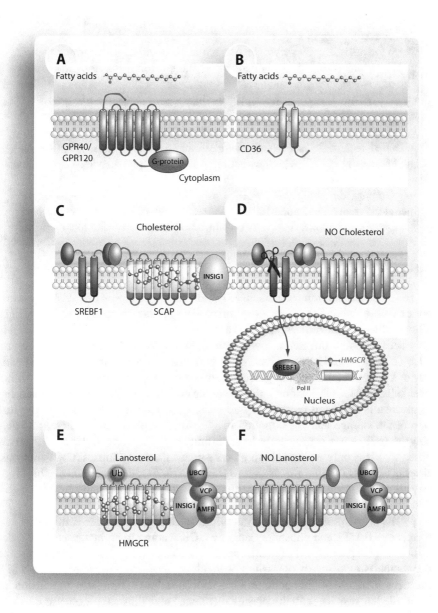

Fig. 3.1 Lipid-sensing mechanisms. Fatty acid detection mechanisms by GPR40 and GPR120 (**a**) and CD36 (**b**). In the presence of cholesterol (**c**), the SCAP-SREBF1 complex binds INSIG proteins and remains anchored to the ER. In the absence of cholesterol (**d**), SCAP-SREBF1 does not bind INSIG1 but moves to the Golgi, where the cytoplasmic tail of SREBF1 is cleaved acting as a transcription factor regulating genes involved in cholesterol synthesis. The ER-embedded enzyme HMGCR catalyzes a rate-limiting step in cholesterol synthesis and is expressed at low cholesterol levels (**e**). When intermediates of the cholesterol biosynthetic pathway, such as lanosterol, are abundant, HMGCR interacts with INSIG proteins leading to HMGCR ubiquitination and degradation (**f**)

Box 3.1 Autophagy

Mechanisms that deliver cytoplasmic molecules of endogenous or exogenous origin to the lysosome for degradation. In **microautophagy** the cargo is directly internalized in small vesicles that fuse with the lysosome, while in **chaperone-mediated autophagy** proteins bearing KFERQ-like motifs are recognized by HSPs (heat-shock proteins) A8 and translocated across the lysosomal membrane. In **macroautophagy** the molecules that should degraded are progressively sequestered within the autophagosome (a double-membrane organelle), which eventually fuses with the lysosome. Lysosomal hydrolases initiate the degradation of the autophagic cargo and recycling of the autophagic products back to the cytoplasm, in order to feed bioenergetics metabolism or repair pathways. All eukaryotic cells exhibit constitutive autophagic flux in physiological conditions, which is essential for the preservation of homeostasis. Autophagic degradation increases in response to a variety of nutritional, hormonal, chemical and physical stresses (Sect. 7.4).

The plasma membrane protein GLUT2 (encoded by the gene *SLC2A2*) is a transporter with a rather low affinity for glucose (Fig. 3.2a). In contrast to other high-affinity glucose transporters, **GLUT2 acts as a true sensor for glucose**, since it is only active at high but not at low physiologic glucose concentrations. Therefore, GLUT2 has a central role in directing the handling of glucose after feeding. At periods of hypoglycemia, hepatic gluconeogenesis increases the glucose levels within liver cells, and GLUT2 exports glucose to the circulation. The intracellular sensing of glucose is mediated by the enzyme GCK (glucokinase) that catalyzes the first step in the storage and consumption of glucose, *i.e.*, glycogen synthesis and glycolysis. GCK has a significantly lower affinity for glucose than the other hexokinases, *i.e.*, it is only active at high glucose concentrations. Thus, **GCK functions, similar to GLUT2, as a glucose sensor** (Fig. 3.2b). At low glucose levels this property allows GCK (in collaboration with GLUT2) to export non-phosphorylated glucose from the liver to the brain and skeletal muscles.

β cells of the pancreas sense systemic glucose levels. Glucose is imported into β cells by GLUT2 and phosphorylated by GCK leading to an increased ATP (adenosine triphosphate)/ADP (adenosine diphosphate) ratio. This depolarizes the membrane *via* closing of potassium channels at the plasma membrane, leads to a transient increase of intracellular Ca^{2+} concentrations, stimulates the fusion of insulin-laden vesicles with the plasma membrane and allows their release into systemic circulation (Fig. 3.2c). In taste buds the detection of high energetic food is mediated by the TAS1R2 (taste receptor, type 1, member 2) in complex with TAS1R3 (Fig. 3.2d). Millimolar concentrations of glucose, fructose or sucrose but also artificial sweeteners, such as saccharine, cyclamate and aspartame, activate the T1R2-T1R3 receptor resulting in the sense "sweet".

Further insight on nutrient sensing systems, including that *via* nuclear receptors (Sect. 3.2), will allow a more integrative view of the molecular reactions of our body

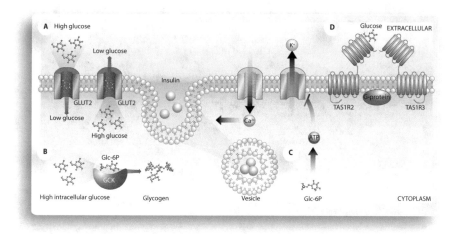

Fig. 3.2 Glucose-sensing mechanisms. Due to its low affinity for glucose the transporter GLUT2 imports glucose only, when it has high concentrations (**a, right**) and exports glucose from hepatocytes into the circulation at hypoglycemic conditions (**left**). The enzyme GCK has low affinity for glucose, *i.e.*, only at high glucose concentrations it produces Glc-6P (glucose-6-phosphate) for the use in glycolysis or glycogen synthesis (**b**). The release of insulin from β cells of the pancreas is a multistep process that involves the phosphorylation of glucose by GCK, subsequent ATP production and ATP-mediated blocking of potassium channels (**c**). A resulting calcium influx facilitates the release of insulin from vesicles into the bloodstream. The heterodimeric oral taste receptors TAS1R2-TAS1R3 bind only high concentrations of glucose, sucrose, fructose and artificial sweeteners and trigger signal transduction through G proteins (**d**)

to dietary molecules. This will not only address the cross-regulation between different nutrient-sensing pathways, but will also incorporate other signaling pathways, such as those controlling cellular growth or mediating chronic inflammation (Sect. 1.5).

3.2   Nuclear Receptors as Nutrient Sensors

Most extracellular signaling molecules, such as growth factors and cytokines, are hydrophilic and cannot pass cellular membranes, *i.e.*, they need to interact with membrane receptors, in order to activate a signal transduction pathway that eventually leads *via* the activation of a transcription factor to changes in gene expression (Sect. 5.2). Thus, **transcription factors serve as sensors of a multitude of cellular perturbations**. In contrast, in case of lipophilic signaling molecules, such as steroid hormones, the signal transduction process is more straightforward, since these compounds can pass cellular membranes and bind directly to nuclear receptors, *i.e.*, to ligand-sensitive transcription factors that are often already located in the nucleus (Fig. 3.3).

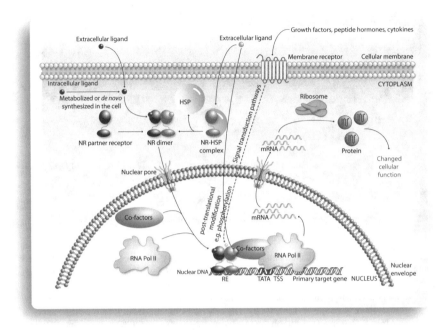

Fig. 3.3 Principles of nuclear receptor signaling. Some nuclear receptors, such as GR and AR, reside in the cytoplasm in a complex with chaperone proteins, such as HSPs, but most nuclear receptors are already located in the nucleus, where they are activated through the binding of their specific lipophilic ligand. The ligand is either of extracellular origin and has passed cellular membranes or is a metabolite that was synthesized inside the cell. After ligand binding, cytoplasmic nuclear receptors dissociate from their chaperones and translocate to the nucleus, where they bind, like the other members of the superfamily, to their specific genomic binding sites, referred to as REs, in the relative vicinity of TSS (transcription start site) regions of their primary target genes. Ligand-activated nuclear receptors interact with nuclear co-factors that build a bridge to the basal transcriptional machinery with Pol II (RNA polymerase II) in its core. This then leads to changes in the mRNA and protein expression of the target genes

The superfamily of nuclear receptors has 48 members in humans and comprises special types of transcription factors that are able to bind and to be activated by small lipophilic molecules called ligands. Many of these nuclear receptor ligands are micro- and macronutrients or their metabolites. This includes the vitamin A derivative retinoic acid (activating RAR (retinoic acid receptor) α, β and γ), fatty acids and other lipids (activating PPARα, δ and γ), 1,25(OH)$_2$D$_3$ (1,25-dihydroxyvitamin D$_3$, activating VDR), oxysterols (activating LXRα and β), bile acids (activating FXR (farnesoid X receptor)) and other hydrophobic food ingredients CAR (constitutively androstane receptor) and PXR (pregnane X receptor). The affinity of these nuclear receptors for their respective ligands ranges between 0.1 nM (for VDR) and more than 1 mM (for PPARs) and reflects the physiological concentrations of the molecules. Thus, **some nuclear receptors represent**

true sensors for micro- and macronutrients. In contrast, other nuclear receptors, such as HNF (hepatocyte nuclear factor) 4α and 4γ, LRH-1 (liver receptor homolog-1), REV-ERB (Reverse-Erb) α and β, ROR (RAR-related orphan receptor) α, β and γ as well as SF-1 (steroidogenic factor 1), bind nutrient derivatives, such as fatty acids, phospholipids, heme and sterols, but this interaction is constitutive and does not represent any sensing process.

All true nutrient sensing nuclear receptors form heterodimers with the sensor for 9-*cis* retinoic acid, RXR (retinoid X receptor) α, β, γ, and bind to specific nucleotide sequences, referred to as REs (response elements). However, other nuclear receptors form homodimers or contact DNA even as monomers. RXR heterodimer complexes permanently locate in the nucleus, *i.e.*, in contrast to GR (glucocorticoid receptor) and AR (androgen receptor) they do not have first to dissociate from chaperone proteins and then to translocate into the nucleus. This indicates that the macro- and micronutrient sensing process *via* nuclear receptors takes place in the nucleus. Thus, **nutrients can act as switches of genes** by inducing a conformational change to the ligand-binding domains of their specific nuclear receptors. This results in the coordinated dissociation of co-repressors and the recruitment of co-activator proteins, in order to enable transcriptional activation of up to 1000 genes per nuclear receptor (Box 3.2).

Box 3.2 Genome-wide nuclear receptor analysis

Different next-generation sequence methods, such as ChIP-seq(chromatin immunoprecipitation followed by sequencing), which were intensively used by the *ENCODE Project* (Box 2.4), have also been applied for the genome-wide analysis of the action of nuclear receptors. The total sum of individual binding sites for an individual nuclear receptor, referred to as its cistrome, collectively for multiple tissues can exceed 20,000 loci. Moreover, transcriptome-wide methods, such as RNA-seq (RNA sequencing), identified, in the sum of all tissues, more than 1000 primary target genes for most nuclear receptors or their specific ligands. Not all of these binding sites and target genes are equally important, but their huge numbers indicate that nuclear receptors and their ligands are involved in the control of more physiological processes than formerly assumed. On many, if not on all of their genomic binding sites nuclear receptors co-locate with other transcription factors, such as SPI1 (spleen focus forming virus proviral integration oncogene, also called PU.1), FOX (forkhead box) A1 or NFκB (nuclear factor κB), that either work as pioneer factors to open the local chromatin structure or to interact with other signal transduction pathways, of which these proteins are the end points. In addition, nuclear receptors do not directly contact DNA on all of their genomic binding sites but can sometimes act as co-factors to other DNA-binding transcription factors.

Members of the nuclear receptor superfamily are involved in the regulation of nearly all physiological processes of our body. They represent a class of transcription factors that can easily and very specifically be regulated by small lipophilic compounds. **Nuclear receptors and their ligands play an important role in the maintenance of homeostasis of our body that represents "health".** The evolutionary oldest and likely still the most important role of nuclear receptors is the regulation of metabolism. There is an interrelationship of lipid metabolism (supplemented by micro- and macronutrients taken up by diet), metabolites and their converting enzymes, such as CYPs (cytochrome P450s), transporters and key representatives of the nuclear receptor superfamily. There are many examples (RAR, CAR, PXR, PPAR, VDR, LXR and FXR, differently color-coded in Fig. 3.4), where a metabolite activates a nuclear receptor, which in turn regulates the expression of the enzyme or transporter handling the metabolite. Nuclear receptor-controlled CYP enzymes have also a central role in receptor ligand inactivation and clearance. These **triangle regulatory circuits are found at several critical positions in lipid metabolism pathways and allow a fine-tuned control on metabolite concentrations and nuclear receptor activity**. This suggests that dietary metabolites are ancestral precursors of endocrine signaling molecules, such as ste-

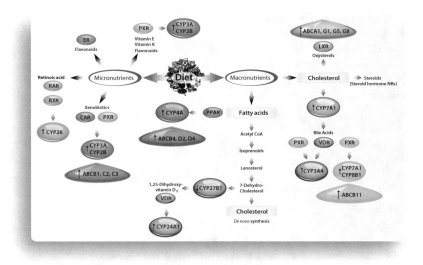

Fig. 3.4 Triangle regulatory circuits of nuclear receptors, their ligands and metabolite handling enzymes and transporters. The interrelationship between micro- and macronutrient metabolism involves enzymes, transporters and nuclear receptors. Only a selected number of metabolites and proteins are shown. There are many examples of triangle relationships (differently color-coded), in which the metabolite regulates its nuclear receptor, the receptor the expression of the metabolite converting enzyme and the enzyme the metabolite levels

roid hormones. In turn this emphasizes the nutrigenomics principle (Sect. 4.4), that **diet is not only a supply for energy but has also important signaling function**.

An immediate implication that followed from understanding the function of nuclear receptors is their potential as therapeutic targets. In fact, nuclear receptor targeting drugs are widely used and commercially successful. For example, bexarotene and alitretinoin (RXRs), fibrates (PPARα), and thiazolidinediones (PPARγ) are already approved drugs for treating cancer, hyperlipidemia and T2D, respectively. Moreover, FXR and LXR agonists are in development for treating NAFLD (non-alcoholic fatty liver disease) and preventing atherosclerosis (Sect. 10.2). However, **natural nuclear receptor ligands that are taken up by healthy diet may avoid any drug treatment**.

3.3   Functions and Actions of PPARs

The different steps in handling fatty acids, such as resorbing them in the intestine, metabolizing them in the liver, burning them in active tissues and collecting their excess for long-term storage in adipose tissue, is coordinated by the three members of the PPAR family (Fig. 3.5). **PPARs are activated by various fatty acids and their derivatives, such as PUFAs, eicosanoids and oxidized phospholipids**. Each PPAR subtype has unique functions, which is based on its distinct tissue distribution. PPARα is expressed predominantly in the liver, heart and kidney. PPARδ is ubiquitously expressed but has most important functions in skeletal muscle (Sect. 1.6), liver and heart. PPARγ is highly expressed in adipose tissue and acts there both as a master regulator of adipogenesis (Sect. 8.2) and a potent modulator of lipid metabolism and insulin sensitivity. Due to alternative splicing and differential promoter usage, there are two PPARγ isoforms, of which PPARγ1 is expressed in many tissues, while the expression of PPARγ2 is restricted to adipose tissue.

During fasting or starvation, PPARα is the primary regulator of the adaptive response in the liver (Fig. 3.5a). This receptor senses the reversed flux of fatty acids and activates a gene network to oxidize fatty acids, in order to generate energy in liver and muscle and to convert fatty acids into a usable energy source, such as ketone bodies during starvation. PPARα is also the molecular target of fibrates, which are widely used drugs that reduce serum triacylglycerol levels through increased fatty acid β-oxidation. In addition, PPARα stimulates the production and secretion of the hepatokine FGF (fibroblast growth factor) 21, which acts as a stress signal to other tissues, in order to adapt to an energy-deprived state in case of fasting. After a meal, PPARα and PPARδ are sensing increasing fatty acid efflux from the liver and start to manage lipid metabolism *via* the promotion of fatty acid β-oxidation and ATP production in mitochondria of skeletal muscles and the heart. Thus, **fatty acids stimulate their breakdown**.

PPARα, PPARδ and PPARγ interfere with the transcription factors NFκB and AP1 (activating protein 1) in macrophages, endothelial cells, epithelial cells and other tissues, causing the attenuation of pro-inflammatory signaling by decreasing

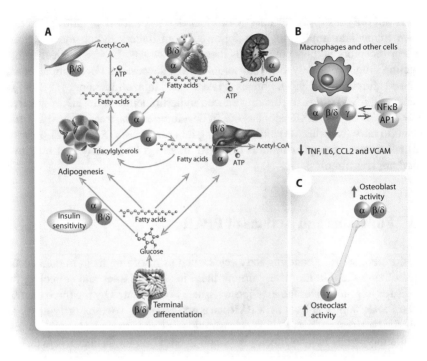

Fig. 3.5 Physiological roles of PPARs. PPARα regulates the expression of enzymes that lead to the mobilization of stored fatty acids in adipose tissue and of fatty acid catabolizing enzymes in the liver, heart and kidney (**a**). PPARδ (also called PPARβ in rodents) is expressed at high levels in the intestine where it mediates the induction of terminal differentiation of epithelium. Activating PPARδ or PPARγ can increase insulin sensitivity. PPARδ regulates the expression of fatty acid catabolizing enzymes in skeletal muscle where released fatty acids are oxidized to generate ATP. PPARγ promotes the differentiation of adipocytes. PPARα, PPARδ and PPARγ can interfere with the transcription factors NFκB and AP1 causing the attenuation of pro-inflammatory signaling (**b**). All tree PPAR subtypes also act in bone (**c**)

the expression of pro-inflammatory cytokines, chemokines and cell adhesion molecules (Fig. 3.5b). For example, PPARα and PPARδ neutralize *via* transrepression the p65 subunit of NFκB and in this way inhibit the expression of NFκB-controlled cytokines, such as TNF, IL1B and IL6. Thus, **PPARs are involved in the control of inflammation** (Sect. 7.2).

PPARδ decreases serum triacylglycerol levels, prevents high-fat diet-induced obesity and increases insulin sensitivity through the regulation of genes encoding fatty acid metabolizing enzymes in skeletal muscle and genes encoding for lipogenic proteins in the liver. Moreover, PPARδ increases serum HDL-cholesterol levels *via* stimulating the expression of the reverse cholesterol transporter ABC (ATP-binding cassette) A1 and APO (apolipoprotein) A1, which is specific for cholesterol efflux (Sect. 10.2). Moreover, the activation of PPARα and PPARδ promotes osteoblast activity in bone, whereas the activation of PPARγ leads to osteoclast stimulation (Fig. 3.5c).

In adipose tissue, PPARγ controls glucose uptake *via* regulating the expression of the *SLC2A4* gene (encoding for GLUT4). Furthermore, PPARγ together with FGF1 mediates adipose tissue remodeling for maintaining metabolic homeostasis during famine. Since high concentrations of circulating fatty acids can cause insulin resistance (Sect. 9.2), enhanced uptake of fatty acids in adipose tissue, the stimulation of the secretion of adiponectin and the inhibition of resistin production *via* PPARγ improve the condition of the disease. Thus, **activating PPARs by specific synthetic ligands (glitazones) can inhibit obesity-related insulin resistance**. However, the increased uptake of fatty acids as well as the enhanced adipogenic capacity in WAT after PPARγ activation is responsible for thiazolidinedione-associated weight gain. Moreover, the thiazolidinedione rosiglitazone was found to increase the risk of heart failure, myocardial infarction and CVD, leading to restricted access in the United States and a market withdrawal in Europe. Nevertheless, the activation PPARγ *via* its natural ligands, such as PUFAs taken up by healthy diet, may be sufficient.

3.4 Integration of Lipid Metabolism by LXRs and FXR

LXRs and FXR are sensors for the cholesterol derivatives oxysterols and bile acids, respectively. These nuclear receptors do not only regulate cholesterol and bile acid metabolism but also have a central role in the integration of sterol, fatty acid and glucose metabolism. LXRα is expressed in tissues with a high metabolic activity, such as liver, WAT and macrophages, whereas LXRβ is found ubiquitously. LXRs, similarly to PPARs, have a large hydrophobic ligand-binding pocket that can bind to a variety of different ligands, such as the oxysterols 24(S)-hydroxycholesterol, 25-hydroxycholesterol, 22(R)-hydroxycholesterol and 24(S),25-epoxycholesterol, at their physiological concentrations. FXR is expressed mainly in the liver, intestine, kidney and adrenal glands. Bile acids, such as chenodeoxycholic acid and cholic acid, are endogenous FXR ligands, but the molecules can also activate the nuclear receptors PXR, CAR and VDR.

LXR is known best for its ability to promote reverse cholesterol transport (Sects. 7.3 and 10.2), *i.e.*, cholesterol delivery from the periphery to the liver for excretion (Figs. 3.6, 7.5 and 10.3). This involves the transfer of cholesterol to APOA1 and pre-β HDLs *via* the transporter protein ABCA1, which is encoded by one of the most prominent LXR target genes. Further important LXR targets are ABCG1 and ABCA1 that promote cholesterol efflux from macrophages, and the intracellular trafficking protein ARL4C (ADP-ribosylation factor-like 4C) that facilitates cholesterol delivery to the plasma membrane. LXR also downregulates the *LDLR* (low-density lipoprotein (LDL) receptor) gene and upregulates the *IDOL* (inducible degrader of LDLR) gene. Thus, **activation of LXR attenuates LDL uptake by macrophages counteracting the pathogenesis of atherosclerosis** (Sect. 10.2).

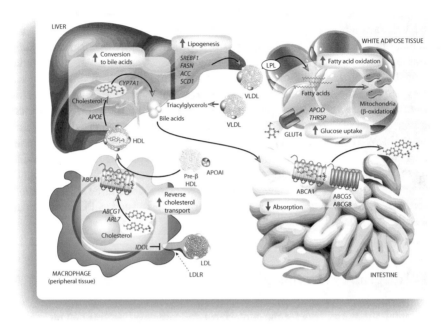

Fig. 3.6 Effects of LXR on metabolism. LXR has effects on multiple metabolic pathways. In macrophages, LXR induces the expression of the genes *IDOL*, *ARL4C*, *ABCA1* and *ABCG1*. In the liver, LXR promotes fatty acid synthesis *via* induction of the transcription factor SREBF1 and its target genes *FASN*, *ACC* and *SCD1*. Triglyceride-rich VLDLs (very low-density lipoproteins) in the liver serve as transporters for lipids to peripheral tissues, including adipose tissue, where the action of LPL liberates fatty acids from VLDLs. In adipose tissue, LXR regulates the expression of *APOD* and *THRSP* and promotes fatty acid β-oxidation and glucose uptake *via* induction of GLUT4. Finally, in the intestine, LXR inhibits cholesterol absorption by inducing the expression of the ABCG5-ABCG8 complex

In the liver, LXR induces the expression of a cluster of apolipoprotein genes including *APOE*, *APOC1*, *APOC2* and *APOC4*, being involved in lipid transport (Sect. 10.3) and catabolism. LXR also upregulates genes encoding for lipid remodeling proteins, such as *PLTP* (phospholipid transfer protein), *CETP* (cholesterol ester transfer protein) and *LPL* (lipoprotein lipase) (Fig. 3.6). Furthermore, LXR induces the expression of several genes that mediate the elongation and the desaturation of fatty acids, which leads to synthesis of long-chain PUFAs, such as ω-3 fatty acids. This increase in long-chain PUFAs results in decreased expression of NFκB target genes *via* epigenetic silencing of their regulatory regions. Long-chain PUFAs also function as substrates for the enzymes that synthesize eicosanoids and other specialized inflammation-resolving lipid mediators, such as resolvins and protectins. Furthermore, LXR induces the gene for the enzyme *LPCAT3* (lysophospholipid acyltransferase 3) mediating the synthesis of phospholipids containing long-chain PUFAs. Thus, **LXR activation decreases ER stress and inflammatory responses** (Sect. 7.5).

A major function of LXR in the liver is the promotion of *de novo* biosynthesis of fatty acids through the stimulation of the transcription factor SREBF1 and the enzymes ACC, FASN and SCD1 (steroyl-CoA desaturase 1). Some of these fatty acids are esterified with cholesterol, in order to avoid toxic levels of free cholesterol. In the intestine, LXR induces the expression of genes encoding the transporters ABCG5 and ABCG8 that mediate the apical efflux of cholesterol from enterocytes. In adipose tissue, LXR also affects glucose metabolism *via* the stimulation of GLUT4 expression. In this tissue, LXR regulates the expression of lipid-binding and metabolic proteins, such as APOD and THRSP (thyroid hormone responsive), and increases fatty acid β-oxidation.

FXR often acts in a complementary or reciprocal way to LXR in the control of lipid metabolism. Since high bile acid concentrations are toxic to cells, a central function for FXR is to control these levels. Bile acids are cholesterol derivatives that facilitate the efficient digestion and absorption of lipids and lipid-soluble vitamins after a meal, but they represent also the major way to eliminate cholesterol from the body. Nevertheless, most bile acids are recycled *via* enterohepatic circulation, *i.e.*, they pass from the intestine back to the liver. FXR controls this bile acid flux *via* modulating their synthesis, modification, absorption and uptake. In the liver, FXR inhibits bile acid synthesis by repressing the expression of the genes *CYP7A1* and *CYP8B1*. FXR stimulates the secretion of FGF19 from the intestine, which then activates FGFR4 (FGF receptor 4) in the liver and in this way provides a complementary mechanism for the feedback inhibition of bile acid synthesis.

In the gall bladder, FXR upregulates the expression of the enzymes SLC27A7 (bile acid-CoA synthase) and BAAT (bile acid-CoA-amino acid N-acetyltransferase), which catalyze the conjugation of bile acid with the amino acids taurine or glycine, and that of the bile salt export pumps ABCB11 and ABCB4. In the intestine, FXR inhibits the absorption of bile salts *via* the downregulation of the apical sodium-dependent bile salt transporter SLC10A2 and promotes the movement of bile salts from the apical to the basolateral membrane of enterocytes *via* the upregulation of FABP6 (ileal fatty acid-binding protein). FXR limits hepatic bile salt levels by the downregulation of the bile acid transporters SLCOs (solute carrier organic anion transporters) A1 and A2. Furthermore, in order to protect the liver from toxicity, FXR induces the expression of the enzymes CYP3A4 and CYP3A11, which hydroxylate bile acids, as well as SULT2A1 (sulfotransferase family 2A, member 1) and UGT2B4 (UDP glucuronosyltransferase 2 family, polypeptide B4), which sulphate and glucuronidate bile acids, respectively. In addition, in the liver FXR reduces lipogenesis *via* the repression of *SREBF1* and *FASN* gene expression, *i.e.*, the receptor decreases triacylglycerol levels. Finally, FXR also influences hepatic carbohydrate metabolism and inhibits gluconeogenesis *via* the downregulation of G6PC (glucose-6-phosphatase) and PCK (phosphoenolpyruvate carboxykinase) 2. Thus, **FXR is an important regulator of lipid metabolism**.

3.5 Coordination of the Immune Response by VDR

The immune system is composed by a multitude of highly specialized cells (Box 2.2) that are all created by a differentiation process of blood cells, referred to as hematopoiesis. Cellular differentiation is controlled by epigenetic mechanisms (Sect. 5.1), in which a number of developmental transcription factors play a key role. **Cells of the immune system have a rapid turnover and are therefore able to show a maximal adaptive response to environmental changes**. Lipid sensing and signaling *via* nuclear receptors has an important role in the differentiation and subtype specification of immune cells, such as T cells, macrophages and DCs (dendritic cells, Sect. 7.1). Importantly, these cells are very mobile and are found in a wide range of subtypes nearly everywhere in our body, *i.e.*, also in metabolic tissues and in disease scenarios, such as obesity (Sect. 8.1). Thus, **macrophages and DCs as well as their precursors, monocytes, coordinate metabolic, inflammatory and general stress-response pathways *via* changes of their transcriptome profile and respective subtype specification**. Nuclear receptors, such as VDR, RAR, LXR and PPAR, have central functions in sensing these endogenous and exogenous stimuli as well as in adapting the respective gene expression profiles of the immune cells. As a representative of all micro- and macronutrient-sensing nuclear receptors, in the following we will focus on VDR and its ligand $1,25(OH)_2D_3$.

Vitamin D_3 and its most abundant metabolite, 25-hydroxyvitamin D_3 (25(OH) D_3), either derive from diet, such as fatty fish, or from endogenous production of vitamin D_3 in response to UV-B exposure of the skin. Since vitamin D and its metabolites can be stored in adipose tissue, there are rather seasonal than daily variations in the vitamin D status of our body. **Worldwide more than one billion people are vitamin D deficient**, *i.e.*, their $25(OH)D_3$ serum levels are below 50 nM. Bone malformations, such as rickets and osteomalacia, are extreme examples of the effects of vitamin D deficiency, but since vitamin D is involved in a broad range of physiological processes, it can increase the risk for various diseases and susceptibility for infections. Living at higher latitudes, *i.e.*, at significant seasonal variations of UV-B exposure, increases the risk of the autoimmune diseases T1D, multiple sclerosis and Crohn's disease.

VDR is expressed in all important cell types of the immune system, *i.e.*, these cells are sensitive to changes in $25(OH)D_3$ serum levels. Importantly, macrophages and DCs express the enzyme CYP27B1 that converts $25(OH)D_3$ into the VDR ligand $1,25(OH)_2D_3$, *i.e.*, in these cells vitamin D can act autocrine or paracrine (Fig. 3.7). Interestingly, while *CYP27B1* expression in the kidneys is negatively regulated by a number of signals, such as Ca^{2+}, parathyroid hormone, phosphate and $1,25(OH)_2D_3$, antigen-presenting cells do not respond to these inhibitory signals but rather further upregulate *CYP27B1* expression after stimulation with cytokines and TLR ligands.

TLRs and other PRRs detect pathogens on the surface of macrophages and initiate an immune response (Sect. 7.2), *e.g.*, against the intracellular bacterium *M. tuberculosis*. Notably, no other infectious disease than tuberculosis had so many human

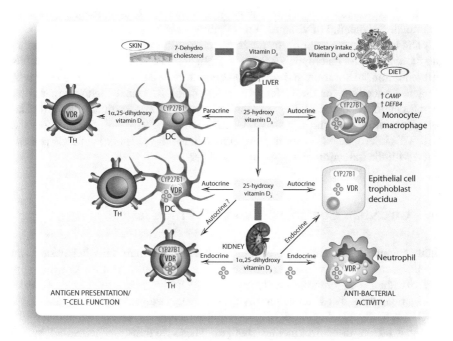

Fig. 3.7 Innate and adaptive immune responses to vitamin D. Macrophages and DCs express the vitamin D-activating enzyme CYP27B1 and VDR can then utilize 25(OH)D$_3$ for autocrine and paracrine responses *via* localized conversion to active 1,25(OH)$_2$D$_3$. In monocytes and macrophages, this promotes the response to infection *via* the anti-bacterial peptides CAMP and DEFB4. 1,25(OH)$_2$D$_3$ inhibits DC maturation and modulates T helper (T$_H$) cell function. Intracrine immune effects of 25(OH)D$_3$ may also occur in *CYP27B1/VDR*-expressing epithelial cells. In contrast, most other cells, such as T$_H$ cells and neutrophils, depend on the circulating levels of 1,25(OH)$_2$D$_3$ that are synthesized by the kidneys, *i.e.*, they are endocrine targets of 1,25(OH)$_2$D$_3$

victims (in total up to one billion). In macrophages, the TLR-triggered increased expression of VDR target genes encoding for anti-microbial peptides, such as *CAMP* (cathelicidin) and *DEFB4* (defensin, beta 4A), efficiently kill intracellular *M. tuberculosis* (Fig. 3.7). This vitamin D-dependent anti-microbial mechanism can explain, why sun or artificial UV-B exposure is efficient in the supportive treatment of tuberculosis, vitamin D deficiency is associated with more aggressive tuberculosis, some variations of the *VDR* gene increase the susceptibility to *M. tuberculosis* infection and humans with dark skin living distant from the equator have an increased susceptibility to tuberculosis infection.

Moreover, vitamin D-induced cytokine production of T cells and monocytes modulate *CAMP* expression. Thus, **the availability of the micronutrient vitamin D is essential for an appropriate response to infections**. Like many other nuclear receptors, VDR can antagonize, *via* transrepression of transcription factors, such as NFAT, AP1 and NFκB, the inflammatory response of immune cells. The resulting decreased expression of cytokines, such as IL2 and IL12, demonstrates the anti-inflammatory potential of vitamin D metabolites.

VDR is a key transcription factor in the differentiation of myeloid progenitors into monocytes and macrophages (Sect. 7.1). In contrast, in DCs vitamin D inhibits differentiation, maturation and immuno-stimulatory capacity *via* the repression of the genes encoding for the different variants of MHC (major histocompatibility complex) and its co-stimulatory molecules CD40, CD80, CD86 and the upregulation of inhibitory molecules, such as CCL22 (chemokine (C-C motif) ligand 22) and IL10. This tolerogenic (*i.e.*, immune tolerance inducing) phenotype of DCs is associated with the induction of T_{REG} cells (Fig. 3.7).

3.6 Circadian Control of Metabolic Processes

Light-sensitive organisms, such as humans, synchronize their daily behavioral and physiological rhythms with the rotation of the Earth around its axis, *i.e.*, they display circadian (*i.e.*, "approximately one day") activity cycles, such as sleep/wake and fasting/eating. These circadian rhythms are under control of a molecular clock, which is a hierarchical network of transcription factors and associated nuclear proteins that adapts to environmental changes. These rhythms are generated by the SCN (suprachiasmatic nucleus) of the hypothalamus (Fig. 3.8a). The SCN is composed of only 15–20,000 neurons, which autonomously oscillate in a 25 h rhythm. *Via* a direct connection with the retina this central clock is adjusted to the daily light-dark cycle. **The SCN is the main driver of circadian fluctuations in blood glucose levels via scheduling food ingestion to the activity phases**. Moreover, the SCN directs the peripheral clocks in all other tissues and cells of our body *via* synchronizing rhythmic food intake. The output of the oscillating system is coordinated physiology *via* the control of various processes in metabolism and behavior. Thus, **the circadian clock is a critical interface between nutrition and homeostasis directing and maintaining proper rhythms in metabolic pathways**.

The core of the circadian clock is a series of transcription-translation feedback loops of the transcription factors ARNTL (aryl hydrocarbon receptor nuclear translocator-like, also called BMAL1) and CLOCK (clock circadian regulator) and their co-repressor proteins. This ARNTL-CLOCK complex activates in a circadian fashion the expression of hundreds of genes both in the brain and in peripheral metabolic tissues, including also the genes *PER1* (period circadian clock 1) and *CRY1* (cryptochrome circadian clock 1). The PER1-CRY1 co-repressor complex inactivates ARNTL-CLOCK, but phosphorylation and ubiquitylation of CRY1 during the night initiates the proteosomal degradation of the repressors and re-activates ARNTL-CLOCK. The genes encoding for the nuclear receptors REV-ERBα and RORα are further targets of ARNTL-CLOCK. REV-ERBα negatively and RORα positively regulates the expression of the *ARNTL* gene, *i.e.*, these nuclear receptors form additional feedback loops in the control of the circadian clock (Fig. 3.8b). In total, the expression of some 10–15% of genes in all organs and tissues displays a circadian rhythm.

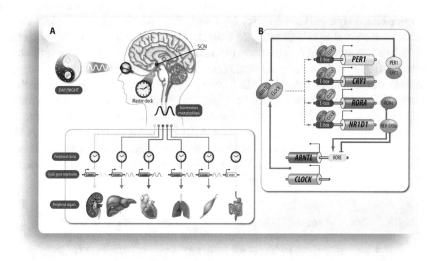

Fig. 3.8 The circadian clock. Electrical and humoral signals from the SCN synchronize phases of circadian clocks in peripheral organs, which then generate time-dependent rhythms in gene expression, metabolism and other physiological activities (**a**). In the feedback loop of the molecular circadian oscillator positive elements, such as the transcription factors ARNTL, CLOCK and ROR, are shown in green, and negative elements, such as PER1, CRY1 and REV-ERBα, in red (**b**). The combined actions of hundreds of ARNTL-CLOCK target genes provide a circadian output in physiology

In the absence of external stimulation, the release of the stress hormone cortisol as well as some peptide hormones, such as thyrotrophin and GH1 (growth hormone 1), follow a circadian rhythm, *i.e.*, the respective endocrine glands are under the influence of circadian clocks. Food intake is organized into distinct meals during the daily cycle, *i.e.*, in the active part of the day energy stores are replenished, while the sleep phase represents a daily period of fasting and mobilization of energy stores. **Bad lifestyle decisions are able to reprogram the circadian clock**. For example, eating at night, artificial light, shift work, travel across time zones and temporal disorganization have disrupted for many humans the alignment between the external light-dark cycle and their internal clock. This is of disadvantage for metabolic health. Longitudinal population studies and clinical investigations both have indicated an association between shift work and diseases, such as T2D, gastrointestinal disorders and cancer that can be modulated by changes in the circadian rhythm. Furthermore, the habit of altering bedtime on weekends, the so-called "social jet lag", is associated with increased body weight.

The circadian clock can be modulated by metabolites, in particular by those representing energetic flux (Box 3.3). For example, the AMP sensor AMPK (Sect. 6.6) connects the internal clock function to the nutrient state *via* phosphorylation and subsequent proteasomal degradation of the ARNTL-CLOCK repressor CRY1.

In parallel, the cyclical activity of ARNTL-CLOCK is modulated by the chromatin modifier KDM (lysine demethylase) 5A (Sect. 5.2), which in turn is linked *via* its co-factors iron and α-ketoglutarate to cellular redox and mitochondrial energetics. The bidirectional interaction between circadian and metabolic signaling is the inhibition of ARNTL-CLOCK by the NAD (nicotinamide adenine dinucleotide)-dependent HDAC (histone deacetylase) SIRT (sirtuin) 1. This represents another feedback control of the circadian clock, since the gene encoding for the critical enzyme for NAD^+ synthesis, *NAMPT* (nicotinamide mononucleotide phosphoribosyltransferase, also known as visfatin), is a direct ARNTL-CLOCK target. Since NAD^+-dependent sirtuins are important regulators of metabolic pathways in response to calorie restriction (Sect. 6.6), the **link between the circadian clock and sirtuin activity has implications for aging**.

Box 3.3 Modulating the circadian clock by metabolic systems
The proteins of the circadian clock are not only expressed in the SCN but also at peripheral sites, such as the liver. Entrainment of the liver clock to feeding involves glucocorticoid signaling, temperature *via* HSF1 (heat shock transcription factor 1) and ADP-ribosylation. In this way, the circadian clock coordinates daily behavioral cycles of sleep-wake and fasting-feeding with anabolic and catabolic processes in the periphery. The central and peripheral clocks are also synchronized *via* post-translational modifications of transcription factors and histones that tune gene expression rhythms into changes of the metabolic state. Therefore, in addition to the transcription-translation-based feedback systems, mammals and other species use NAD^+ oscillation, redox flux, ATP availability and mitochondrial function, in order to influence acetylation and methylation reactions. For example, the redox-based clock represents oscillations in the redox state of the family of peroxiredoxin antioxidant enzymes that rhythmically anticipate the generation of ROS. Moreover, NAD^+ is an electron shuttle in oxidoreductase reactions and also acts as a co-factor in HDAC and ADP-ribosylation modifications (Sect. 5.2).

Additional Readings

Carlberg C, Molnár F (2016) Mechanisms of gene regulation. Springer Textbook ISBN: 978-94-017-7740-7

Challet E (2019) The circadian regulation of food intake. Nat Rev Endocrinol 15:393–405

Efeyan A, Comb WC, Sabatini DM (2015) Nutrient-sensing mechanisms and pathways. Nature 517:302–310

Greco CM, Sassone-Corsi P (2019) Circadian blueprint of metabolic pathways in the brain. Nat Rev Neurosci 20:71–82

Reinke H, Asher G (2019) Crosstalk between metabolism and circadian clocks. Nat Rev Mol Cell Biol 20:227–241

Chapter 4
Interference of the Human Genome with Nutrients

Abstract This chapter will discuss the complex relation between our environment, diet and genome. Genes influence our response to diet, while nutrients, or the lack of them, can affect gene expression. More than 90% of our genes have not changed since the life in the stone ages, where food availability meant survival. In this context, the molecular basis for the recent adaption of our genome to environmental changes, such as less UV-B exposure after migrating north, and dietary opportunities due to dairy farming, such as lactose tolerance, will be discussed. The majority of trait-associated variants of our genome are located outside of protein-coding regions, *e.g.*, often they are regulatory SNPs within transcription factor binding sites. Nutrigenomics has taken up many elements from molecular biology and next-generation sequencing technologies for investigating the effects of food on the level of our epigenome, genome, transcriptome, proteome and metabolome. These methods can be applied for comprehensive assessment of individuals, such as in iPOP (integrated personal omics profile)-style analyses. The respective datasets are the basis for the optimization of personalized nutrition, preserving health *via* the prevention of nutrition-related diseases.

Keywords Immunity · Skin color · Positive selection · Lactose tolerance · Expression quantitative trait locus · Regulatory SNPs · Nutrigenomics · Omics technologies · Integrative personal omics profile · Polygenic risk score · Personalized nutrition

4.1 Human Genetic Adaptions

Changes in environment and nutrition have been a major driver of human evolution and may have been the main factor that enabled *Homo sapiens* to survive and progress (Sect. 1.1). Humans have spread from Africa around the world, experienced an ice age, domesticated hundreds of plant species and more than a dozen animals for the start of agriculture and dairy farming (Sect. 2.1). Thus, **during the migration of the past 50,000 years, selective pressures in local environments in combination**

© Springer Nature Switzerland AG 2020

C. Carlberg et al., *Nutrigenomics: How Science Works*,

https://doi.org/10.1007/978-3-030-36948-4_4

with random genetic drifts resulted in population-specific genetic adaptations. In parallel, humans significantly increased in population density, *i.e.*, they shifted from small groups of mobile hunters and gatherers to permanent settlement of larger numbers of families in villages. This human-modified habitat favored vector insects, such as mosquitos, and involved close proximity to domesticated animals, such as chicken and pigs, both of which act as reservoirs of zoonotic pathogens, such as viruses, bacteria and parasites. Thus, **compared to hunters and gatherers the burden of infectious diseases in agricultural societies increased drastically**. The exposure to novel pathogens served as strong challenges for the immune system and selective pressures in recent human evolution.

Interestingly, the admixture of *Homo sapiens* with archaic *homini*, such as Neanderthals and Denisovans, who already had adapted genetically to environmental conditions of Eurasia for more than 300,000 years, led to the transfer of gene variations and improved the fitness for survival outside of Africa. Although in total only 1–2% of the genome of today's populations in Europe and Asia relates to archaic *homini*, some of these variants cluster to larger haplotype blocks. Due to this adaptive introgression process alleles of Neanderthal origin are associated with traits of medical relevance, such as the risk for the autoimmune disease lupus erythematosus, biliary cirrhosis and Crohn's disease and T2D (Sect. 2.2). Thus, **seasonal variations and exposure to cold had made the Neanderthal's immune system more robust, the underlying gene variants of which improved the adaption of *Homo sapiens* to environmental conditions of Eurasia**.

A very obvious phenotypic difference between today's human populations is the color of skin, hair and eyes. These traits largely influence our appearance and may also have impact on reproductive success. Skin is our largest organ and it mediates direct interaction with the environment, such as absorption of UV radiation, tactile sensitivity, detection of pain and thermoregulation. People that live close to the equator or at high altitude, such as in the Himalayas or the Andes, have the darkest skin, while on the northern hemisphere at higher latitude lighter skin types are observed. **Before the migration out of Africa the skin of *Homo sapiens* was dark**, because some 2 million years earlier their ancestors turned dark when they lost most of their body hair, in order to better regulate their body temperature *via* sweating during endurance physical activity (Sect. 1.6). Permanent dark skin better protects against the deleterious effects of solar UV-B radiation, such as sunburns and skin cancer, and may prevent the degradation of the photosensitive B vitamin folate.

Variations in genes affecting pigmentation represent a key example of adaption to the environment in Europe and Asia. In the process of melanogenesis the amino acids phenylalanine, tyrosine and cysteine are converted to melanin. Variations in genes, such as *SLC24A5*, *SLC45A2*, *OCA2* (OCA2 melanosomal transmembrane protein), *TYR* (tyrosinase), *MC1R* (melanocortin 1 receptor), *IRF4* (interferon regulatory factor 4), *DCT* (dopachrome tautomerase), *CTNS* (cystinosin, lysosomal cystine transporter) and *MYO5A* (myosin VA), encoding for the enzymes of this pathway as well as for ion channels in melanocytes or transport molecules involved in melanosome maturation and export can result in light hair, light skin and blue eyes, as it is typical in northern European populations. The amount of eumelanin (black-brown) and pheomelanin (yellowish-redish) production within melano-some granules affects skin and hair color after they have been transferred from melanocytes to keratinocytes or hair shafts, respectively. In contrast, melanocytes within the iris keep their melanosomes and eye color depends on the ratio of phe-omelanin to eumelanin (14:1 for blue eyes and 1:1 for brown eyes). Furthermore, a variant of the *EDAR* (ectodysplasin A receptor) gene, which is required for the development of hair, teeth and other ectodermal tissues, is associated with increased hair thickness, a higher number of sweat glands and shovel-shaped incisors. This variant is dominant in Asian populations.

4.2 Genetic Adaption to Dietary Changes

Humans were the only species that learned some 1 million years ago to use fire for cooking raw food and thereby created a safer and more easily digestible diet. Together with the omnivorous choice of diet, the advantage of cooking increased the energy yield of the meals and allowed the enlargement of glucose-demanding brains. Moreover, like other plant eating species, humans evolved receptors for sensing sweet taste (Sect. 3.1), in order to detect the most energy-rich diet, although **this initial survival instinct nowadays causes overweight and obesity**.

Our diet is majorly composed of starch from grain flour, rice or potatoes (Sect. 1.1). The polysaccharide starch is digested to glucose by enzymes of the *AMY* (amy-lase) gene family, which in some species, including us, are expressed both in saliva (*AMY1*) and pancreas (*AMY2A* and *AMY2B*). In agricultural societies with starch-rich diets individuals tend to have higher copy numbers of the *AMY* genes than hunters and gatherers with low starch consumption. For example, in Japan large amounts of rice and starch from other sources are consumed reflecting many copies of the *AMY1* gene, whereas in the genetically closely related Siberian Yakut popula-tion (primarily eating fish and meat) significantly less *AMY1* copies are found. In general, in contrast to archaic *homini* today's humans have up to 20 copies of the *AMY1* gene, which causes higher levels of salivary AMY protein expression. This leads to better digestion of starchy foods as well as to a sweet sensation in the mouth. **The amplification of the *AMY1* gene is an example of positive evolution-ary selection** (Box 2.1) underlining the long and continuing importance of these staples in our diet. Interestingly, when wolves became dogs they adapted to a new source of food, which were starch-rich leftovers of human diet, and also got multi-ple copies of the *AMY* genes.

The genes of the *ADH* (alcohol dehydrogenase) cluster encode for ethanol metabolizing enzymes and are another example of a gene locus that was positively selected when agriculture made the production of fermented alcoholic beverages easy. These examples suggest that **the transition to new diet sources after the advent of agriculture and the colonization of new habitats have been a major factor for the selection of human genes**. Additional examples of genes that were positively selected due to dietary changes are *ADAMTS* (ADAM metallopeptidase with thrombospondin motif) *19* and *20*, *APEH* (N-acyla(minoacyl-peptide hydro-lase), *PLAU* (plasminogen activator, urokinase) and *UBR1* (ubiquitin protein ligase E3 component n-recognin 1), which encode for enzymes related to protein metabolism. In addition, there are population-specific examples of variants within genes involved in metabolizing mannose (*MAN2A1* (mannosidase, alpha, class 2A, member 1) in West Africa and East Asia), sucrose (*SI* (sucrase-isomaltase) in East Asia) and fatty acids (*SLC27A4* and *PPARD* in Europe, *SLC25A20* in East Asia, *NCOA1* (nuclear receptor co-activator 1) in West Africa and *LEPR* (leptin receptor) in East Asia). Thus, **human populations have genetically adapted to their traditional diet, in order to use best their local resources**.

The probably most prominent example of the adaption of our genome to dietary changes is the use of fresh milk from infanthood to adulthood, referred to as lactase persistence. The disaccharide lactose is the main carbohydrate in milk and is a major energy source for most infant mammals. Lactose is digested into glucose and galactose by the intestinal enzyme LCT (lactase). Lactase non-persistence, also referred to as lactose intolerance, is an autosomal recessive trait that is characterized by diminished expression of the *LCT* gene after weaning, *i.e.*, older children and adults do not express the enzyme anymore. Lactose intolerance was the default genetic setup of early humans (as well as in most other mammals), probably to avoid competition for breast milk between newborns and older children or even adults. Lactose intolerant individuals who consume lactose can have intestinal symptoms, such as bloating, flatulence, cramps, nausea and diarrhea, which may result in nutritional loss even beyond not receiving direct energetic benefits from lactose digestion. Lactose intolerance is still present in some 65% of the global human population, since the new variant of the *LCT* gene has emerged only within the last 5000 years after the domestication of cattle, sheep and goats and the start of dairy farming in Europe. Lactose tolerance is found also in livestock raising populations from Africa and Western Asia, but is almost completely absent elsewhere. Milk drinking created one of the strongest presently known selection pressures on our genome that drove alleles for lactose tolerance to high frequency. In the past, children's mortality rates were very high, since after weaning the loss of immune input from breast milk was associated with multiple exposures to diarrheal diseases. As milk is a perfect source of carbohydrates (primarily lactose), fat and calcium, the ability to use it as a reliable dietary source provides an enormous advantage for survival. Thus, **mutations leading to lactase persistence belong to the most strongly selected genetic variations of all episodes of positive selection in humans**.

The SNPs associated with lactase persistence are located approximately 14 kb and 22 kb upstream of the TSS of the *LCT* gene within introns 13 and 9 of the *MCM6* (minichromosome maintenance type 6) gene (Fig. 4.1). The T/T allele at position −13,910 relative to the TSS of the *LCT* gene (rs4988235) binds the transcription factor POU2F1 (POU class 2 homeobox 1) with higher affinity than the C/T or the C/C variant. Also the transcription factors GATA6 (GATA binding protein 6), CDX2 (caudal type homeobox 2), HNF3A and HNF4A are associated with this regulatory region. As expected, the respective SNPs in the Neanderthal's

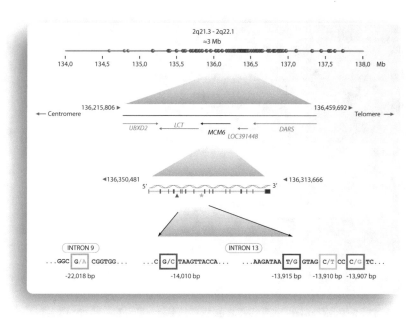

Fig. 4.1 Map of the genomic region of the genes _LCT_ and _MCM6_. Location of the SNPs responsible for lactose tolerance within introns 9 and 13 of the _MCM6_ gene in African and European populations

genome indicated that ancient human populations were lactose intolerant. Furthermore, the larger genomic region around the _LCT_ gene demonstrates significant difference in the size of the respective haplotype blocks between lactose tolerant Europeans (more than 1 Mb) and non-Europeans. This reflects **strong selection for the lactase persistence allele in particular in the northern European population**.

4.3   Regulatory SNPs and Quantitative Traits

GWAS analysis has indicated that some 88% of trait-associated variants are located outside of protein-coding regions of the human genome (Sect. 2.4). The SNP rs4988235 upstream of the _LCT_ gene (Sect. 4.2) represents a master example of a regulatory variant being equally important to SNPs affecting a protein-coding region in determining disease risks and traits (Fig. 4.2). The functional characterization of regulatory SNPs, such as the identification of transcription factor binding to the variant genomic region, can suggest possible therapeutic interventions, _e.g._, when the respective transcription factor is "druggable" (_i.e._, there is a synthetic or

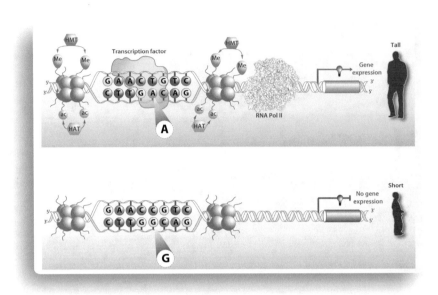

Fig. 4.2 The basis of human trait variation. Small variations within the DNA binding site for a transcription factor can facilitate and even enhance the association of this protein, such as the A (**top**), or inhibit its binding, when it is a G (**bottom**). The binding of the transcription factor influences the local chromatin structure *via* the activation of chromatin modifiers, such as HATs (histone acetyltransferases) and/or HMTs (histone methyltransferases), eventually leading to the activation of Pol II and the transcription of the respective gene (Sect. 5.2). This may have a positive effect on the trait of interest, such as body height. In contrast, when the transcription factor does not bind, the respective genomic region remains inactive and the gene is not transcribed. This may have a negative effect on the studied trait

natural compound that modulates its activity). Gene regulatory events that are related to regulatory SNPs do not only depend on the sequence of the respective genomic site but also on its accessibility within chromatin (Sect. 5.1). **This emphasizes the impact of epigenomics on regulatory variation**.

In contrast to the genome, which is identical in all 400 tissues and cell types of an individual, the epigenome and consequently the expression of genes depends on the individual tissue and the signals that it is exposed to, *i.e.*, it represents the dynamic state of the cell. The next-generation sequencing method RNA-seq in combination with SNP information is the basis of eQTL (expression quantitative trait locus) mapping. This approach allows the functional assessment of a genetic variant on the level of the transcriptome. In general, functional genetic variants can modulate various steps in the process of gene expression from genes *via* mRNAs to active proteins, which are

- changes of the affinity of transcription factors for their genomic binding sites within promoters and enhancers
- a disruption of chromatin interactions

- the modulation of the functionality of ncRNAs
- the induction of alternative splicing
- an alternation in the post-translational modification pattern of proteins.

The effect size of functional SNPs can vary a lot and depends on the affected regulatory process and its epigenomic context, *i.e.*, it is difficult to predict. At present, **changes in the association of transcription factors due to variants in their specific binding sites are the best-understood types of regulatory SNPs** (see examples in Box 4.1).

Box 4.1: Regulatory SNPs with Impact on Obesity and CVD

There are millions of DNA-binding sites for the approximately 1600 sequence-specific transcription factors that are encoded by the human genome. Dependent on their function and position, the regions where these binding sites are clustering, are called promoters, enhancers or insulators.

Example 1: The SNP rs1421085 is located within an intron of the *FTO* (fat mass and obesity associated) gene and shows the most prominent OR for the trait obesity (Sect. 8.5). Its T-to-C variation disrupts the binding site of the transcription factor ARID5B (AT-rich interaction domain 5B), which regulates during early adipocyte differentiation the expression of the genes *IRX* (iroquois homeobox) *3* and *5*. This results in a shift from energy-consuming beige adipocytes to energy-storing white adipocytes (Sect. 8.2), *i.e.*, in a drastic reduction in mitochondrial thermogenesis and in an increase in lipid storage. Knockdown of *IRX3* or *IRX5* or repair of the ARID5B binding site by CRISPR-Cas9 editing restored *IRX3* and *IRX5* repression, activated adipocyte browning and restored thermogenesis. Thus, **not the *FTO* gene but its neighboring genes *IRX3* and *IRX5* functionally explain the prominent effect of SNP rs1421085 on obesity risk.**

Example 2: Myocardial infarction and plasma levels of LDL-cholesterol were are strongly associated with SNP rs12740374 (Sect. 10.3), for which eQTL analysis indicated most significant association with the *SORT1* (sortilin 1) gene. The more active minor allele created a binding site for the transcription factor CEBP (CCAAT/enhancer binding protein), *i.e.*, rs12740374 is a gain-of-function regulatory SNP. **Genomic approaches confirmed *SORT1* as a novel lipid-regulating gene** and its pathway as a target for potential therapeutic interventions.

Results of the projects *ENCODE*, *FANTOM5* and *Roadmap Epigenomics* (Box 2.4) created a unique genome-wide resource that helps to characterize regulatory SNPs on the level of

- post-translational histone modifications, such as methylations and acetylations at various positions of the histones H3 and H4, indicating active and repressed enhancer and promoter regions

- chromatin accessibility
- genomic association of more than 100 transcription factors
- DNA methylation indicating inactive genomic binding profiles
- the non-coding transcriptome of hundreds of cell lines and primary tissues.

The database *RegulomeDB* (http://regulomedb.org) combines information on chromatin state, transcriptional regulator binding and eQTLs allowing the identification and interpretation of functional DNA elements at the sites of regulatory variants.

4.4 Definition of Nutrigenomics

Our diet is a complex mixture of biologically active molecules, some of which can

- have a direct effect on gene expression (Fig. 4.3A)
- modulate, after being metabolized, the activity of a transcription factor or chromatin modifier (Fig. 4.3B)
- stimulate a signal transduction pathway that ends with the induction of a transcription factor (Fig. 4.3C).

Nutrigenomics aims to describe, characterize and integrate these interactions between diet and gene expression genome-wide. The results of these investigations lead to an improved understanding of how nutrition influences metabolic pathways and homeostatic control. This nutrition-triggered regulation may be disturbed in the early phase of a diet-related disease, such as T2D. **When individuals are classified according to the interplay of their lifestyle, metabolic pathways and genetic variation, the molecular insight based on nutrigenomics studies can suggest tailored diets, referred to as personalized nutrition, for early therapeutic intervention**. For example, individuals that are genetically at risk for T2D, but not yet in a prediabetic state, may be recommended a customized diet, in order to avoid developing the disease (Sect. 4.5). Using diet for a specific therapy dissolves the distinction between food and drugs as well as the definition of health and disease. Thus, **the best advice for healthy eating may result in a more individualized lifestyle**.

Both nutrigenomics and pharmacogenomics have some similarities concerning concepts and methodological approaches, *i.e.*, both use high-throughput omics technologies. However, in pharmacogenomics the effects of a single clearly defined compound (a drug) of a precise concentration and a specific target are investigated, whereas nutrigenomics faces the complexity and variability of diets and nutrients. However, some nutritional compounds can reach up to mM concentrations without becoming toxic, while most drugs act at clearly lower concentrations.

From the perspective of nutrigenomics, food is a collection of dietary signals that are detected by cellular sensors, such as membrane proteins and nuclear receptors (Sects. 3.1 and 3.2), leading to changes in the epigenome, transcriptome, proteome and metabolome, *i.e.*, the complete set of all chromatin modifications, mRNA mol-

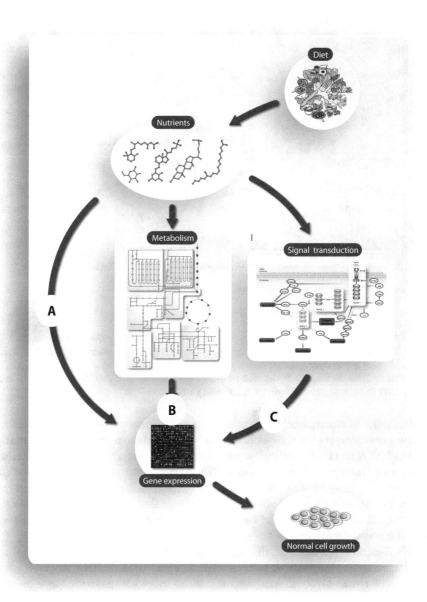

Fig. 4.3 Basis of nutrigenomics. Nutrigenomics seeks to provide a molecular understanding for how dietary nutrients affect health by altering the expression of a larger set of genes. These nutritional compounds have been shown to alter gene expression in a number of ways. For example, they may (**A**) act as direct ligands for transcription factors, (**B**) be modulators for transcription factors or chromatin modifiers after a chemical conversion in primary or secondary metabolic pathways or (**C**) serve as activators of signal transduction pathways that end with the activation of a transcription factor. All three activation pathways modulate physiological effects, such as cellular growth

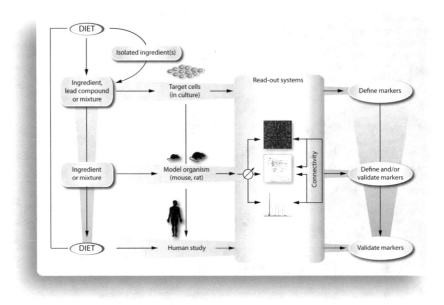

Fig. 4.4 Applications of omics technologies in nutrition research. The effect of food or its ingredients are studied either in cell culture, animal models or human intervention studies. In all cases, samples are analyzed *via* the use of omics technologies on the level of the transcriptome, proteome or metabolome. The integration of these large-scale datasets results in the definition of biomarkers that can be validated *via* complementary studies

ecules, proteins and metabolites in a cell or in a biological sample. Thus, **individual dietary components (as well as food as a whole) induce a pattern in chromatin accessibility, gene expression, protein expression and metabolite production that can be interpreted as "signatures" of the respective nutritional compound**. These dietary signatures can be investigated *in vitro*, such as in cell lines representing metabolic organs, or *in vivo*, such as in rodent model organisms (Fig. 4.4). However, **the most meaningful results will be obtained from intervention studies with human subjects**.

Nutrigenomic technologies that are based on the use of next-generation sequencing methods, *i.e.*, the massive parallel sequencing of DNA or RNA molecules, are presently far more advanced than proteomic and metabolomic methods. Nevertheless, proteome and metabolome data allow more precise measures of a physiological state. At present, the method liquid chromatography mass spectrometry (LC-MS/MS) analysis identifies up to 5000 of the most abundantly expressed proteins. The analysis of metabolites is performed either targeted via GC-MS (gas chromatography-mass spectrometry) and NMR (nuclear magnetic resonance) spectroscopy for a few hundred metabolites or untargeted via LC-MS/MS, which allows the detection of several thousand different molecules. Using the metabolome to characterize individuals has become a powerful tool in nutrition research. Metabotyping describes groups of individuals with similar metabolic profile. This grouping of individuals has the potential to identify optimal treatment strategies for each metabotype.

A central point in nutrigenomic methodology is the integration of the data that are obtained from transcriptomic, proteomic and metabolomic profiling with specific nutrients or diets (Sect. 4.5). **Ideally, this results in the identification of biomarkers that can serve as early warning for nutrient-induced changes to homeostasis**, such as the development of prediabetes (Chap. 9). However, the tight connection of metabolic pathways makes it difficult to achieve specific responses to a treatment with one compound, such as natural or synthetic nuclear receptor ligand, without provoking side effects through compensatory or complementary responses from other pathways. Nevertheless, future studies that will be performed in a safe and clever way in humans (rather than in rodents) should provide a level of understanding that will lead to more specific ligands and identification of new target genes regulated by dietary components. Moreover, a better understanding of the links between circadian biology and metabolism (Sect. 3.6) will allow tailoring preventive interventions and therapies. Thus, **nutrigenomic investigations can provide basic insight on the interaction between nutrition and our genome but can also serve as a diagnostic and potentially therapeutic tool**.

4.5 Personal Omics Profiles

The field of personalized medicine/nutrition is advancing due to the rapid development of omics technologies. For a comprehensive view on health and disease, *i.e.*, when acknowledging the entire complexity of biological processes, these technologies need to be integrated. For example, for a deep understanding of obesity and T2D deep analyses and longitudinal profiling are necessary. A proof-of-principle investigation demonstrating the potential of next-generation technologies had been provided by the iPOP analysis of one individual (Fig. 4.5). The iPOP study included whole genome sequencing and more than 20 times over a period of 14 months sampling was carried out for mRNA and miRNA expression in PBMCs (peripheral blood mononuclear cells), proteome profile in PBMCs and in serum as well as the metabolome and auto-antibodyome in blood plasma. These molecular datasets were complemented by medical lab tests for regular blood biomarkers. Interestingly, both RNA-seq transcriptomics and LC-MS/MS proteomics demonstrated that the genes involved in insulin signaling and response were downregulated during a respiratory syncytial virus infection, which paralleled with increased blood glucose levels up to T2D levels (Sect. 9.1). The integrative profile monitored both gradual trend changes as well as spike changes in particular at the onset of each physiological state adjustment. Thus, **the iPOP analysis allowed a most comprehensive view on the biological pathways that led to the onset of hyperglycemia of the study subject**. Importantly, hyperglycemia was detected in a very early stage, so that it could be effectively controlled and reversed by changing diet and intensified physical activity of the individual.

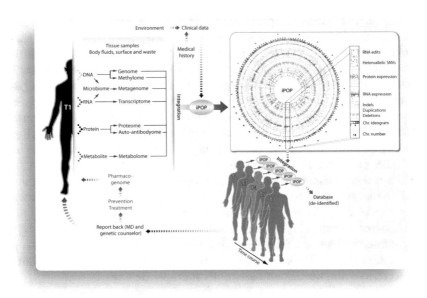

Fig. 4.5 Implementation of iPOP for personalized medicine. Tissue samples (*e.g.*, PBMCs) of an iPOP participant are collected at time points T1 to Tn, while diet, exercise, medical history and present clinical data are also recorded. The results of the iPOP analysis can be monitored by Circos plots (**right**), in which DNA (outer ring), RNA (middle ring) and protein (inner ring) data match to chromosome position. The data may be reported back to genetic counselors and/or medical practitioners, in order to allow most rational choices for prevention and/or treatment, which may be matched with pharmacogenetic data. (Data are based on Chen et al., Cell **148**, 1293–1307 (2012))

The central aim of iPOP-style analyses are early detection of diseases or their prevention as well as their efficient intervention and therapy. A follow-up iPOP-style study involving 23 individuals demonstrated that inflammatory signatures affecting metabolic pathways during weight gain did not to return to baseline after subsequent weight loss. In addition, a longitudinal monitoring of 108 individuals over 9 months on the level of their genome, proteome and metabolome in relation to clinical data connected molecular measurements with physiology and disease. The Genome Aggregation Database (https://gnomad.broadinstitute.org) collects omics data of healthy individuals from a wide variety of large-scale sequencing projects and makes them available to the scientific community.

Disease risk is primarily based on genetic susceptibility, environmental exposures and lifestyle factors. For example, the relative contribution of genetic susceptibility to disease predisposition is quantified by the heritability, *i.e.*, the proportion of phenotypic variation that can be explained by genetic variation, of the disease in a given population (Sect. 2.4). Thus, **most diseases have a polygenic risk score,**

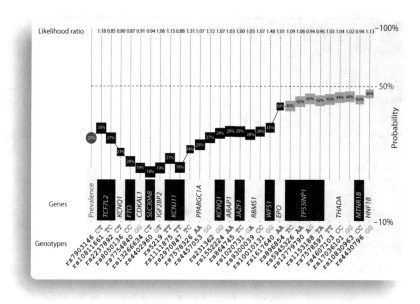

Fig. 4.6 The potential of personalized medicine. The riskogram illustrates how the iPOP study volunteer's post-test probability of T2D was calculated on the basis of 28 independent SNPs (Sect. 9.6). The likelihood ratio is indicated on top, the central graph displays the post-test probability, while the associated genes, SNPs and the subject's genotypes are shown on the bottom. (Data are based on Chen et al., Cell **148**, 1293–1307 (2012))

which is mostly calculated as a weighted sum of the number of risk alleles carried by an individual (Fig. 4.6). Accordingly, a "riskogram" takes age, gender and ethnicity as well as multiple independent disease-associated SNPs into account, in order to determine the subject's likelihood of developing a disease. The original iPOP study calculated for the investigated subject a previously unexpected increased risk to develop hyperglycemia, which in fact was confirmed experimentally after a viral infection.

Future personalized health care as well as the emerging field of personalized nutrition will benefit from a longitudinal recording of physiological and biochemical parameters, such a blood glucose levels, physical activity and blood pressure, *via* wearable biosensors (Box 4.2). These basic data in combination with personal genomic information and tailored iPOP-style studies can be applied to monitor any disease or physiological state changes of interest. The integrative profile of iPOP-style studies is modular and allows the addition of further omics information, such as epigenome-wide data and the microbiome of skin, oropharynx, nasopharynx, stomach, intestinal mucosa or urine as well as quantifiable environmental factors. Thus, **iPOP-style analyses may become central to nutrigenomic projects**.

Box 4.2: Wearable Biosensors
For a longitudinal description of physiological information of an individual, wearable biosensors perform a dynamic, non-invasive monitoring of biomarkers. The first generation of biosensors is based on physical measurements, *i.e.*, they display mobility and vital signs, such as steps, heart rate and sleep quality. More advanced biosensors use optical and electrochemical processes for tracking metabolites (*e.g.*, glucose or lactate), electrolytes (*e.g.*, sodium, potassium or calcium), bacteria and hormones in biofluids like sweat, intestinal fluids, saliva and tears. A prime focus is the sensing of glucose. The new generation sensors are miniaturized and combined with flexible materials, in order to improve wearability. The sensors are composed of a bioreceptor, such as an enzyme, antibody or DNA, and a physico-chemical transducer that translates the biorecognition event into a useful signal, which is transmitted wireless to a mobile device, such as a smartphone

Whole genome sequence information is already available for some 1.5 million individuals and soon it will be common for everyone to have his/her genome sequenced. The rapid development of omics technologies in combination with decreasing costs will allow collecting iPOP-style datasets on many individuals. The integration of such data will allow further exploring the relationship between human genetic variations and complex diseases and respective traits. In particular, **the systematic exploration of epigenomics** (Chap. 5) **will provide critical insights into disease susceptibility**. The ability to stratify individuals according to their genotype will make clinical trials more efficient by enrolling a lower number of subjects with an anticipated larger effect when personalizing the intervention. Diseases, such as T2D (Chap. 9), will be classified into subphenotypes based on the genotype and the dynamic reply of the individual, *e.g.*, in response to a personalized diet. **This will allow using diet for preserving health and for an improved personalized therapy**, most likely in combination with synthetic drugs, in case of disease.

Additional Readings

James WPT, Johnson RJ, Speakman JR, Wallace DC, Frühbeck G, Iversen PO, Stover PJ (2019) Nutrition and its role in human evolution. J Intern Med 285:533–549

Karczewski KJ, Snyder MP (2018) Integrative omics for health and disease. Nat Rev Genet 19:299–310

Kim J, Campbell AS, de Avila BE, Wang J (2019) Wearable biosensors for healthcare monitoring. Nat Biotechnol 37:389–406

Marciniak S, Perry GH (2017) Harnessing ancient genomes to study the history of human adaptation. Nat Rev Genet 18:659–674

Pavan WJ, Sturm RA (2019) The genetics of human skin and hair pigmentation. Annu Rev Genomics Hum Genet 20:41–72

Prohaska A, Racimo F, Schork AJ, Sikora M, Stern AJ, Ilardo M, Allentoft ME, Folkersen L, Buil A, Moreno-Mayar JV et al (2019) Human disease variation in the light of population genomics. Cell 177:115–131

Scheinfeldt LB, Tishkoff SA (2013) Recent human adaptation: genomic approaches, interpretation and insights. Nat Rev Genet 14:692–702

Chapter 5
Nutritional Epigenetics

Abstract In this chapter, we will present nutritional epigenetics as a subdiscipline of nutrigenomics and describe how dietary compounds affect our epigenome. Different epigenetic mechanisms, such as post-translational histone modifications and DNA methylation, process information provided by dietary molecules. Accordingly, many chromatin modifiers use intermediary metabolites, such as acetyl-CoA, α-ketoglutarate, NAD^+ or ATP, as co-substrates and/or co-factors. Thus, these enzymes act as sensors for the nutritional status of our tissues and cell types leaving respective marks on their epigenome. Prenatal supplementation in mice as well as natural human experiments provide insight into the concepts of epigenetic programming during embryogenesis and epigenetic drift during adult life. This may explain some of the susceptibility for complex metabolic diseases, such as T2D.

Keywords Epigenome · Chromatin · DNA methylation · Histone modifications · Chromatin modifiers · Intermediary metabolism · Acetyl-CoA · NAD^+ · Folate · Methylenetetrahydrofolate reductase · Agouti mice · Epigenetic programming · Epigenetic epidemiology · Epigenetic drift

5.1   Epigenetic Mechanisms

Chromatin is the 3D complex of genomic DNA and histone proteins (Box 5.1). It is subdivided into less densely packed euchromatin, which is easily accessible to transcription factors (Fig. 5.1, top left), and compact heterochromatin, which represents a functionally repressed state (Fig. 5.1, top right). **Epigenetics is the study of functionally relevant modifications of chromatin that do not involve a change in the comprised genome sequence**. Epigenetic alternations may remain stable during cell divisions and can last for multiple (cell) generations. The best example is the process of cellular differentiation, where due to epigenetic changes formerly totipotent stem cells become various pluripotent cell lines of the embryo, which in turn are the precursors of terminally differentiated cells. The main epigenetic mechanisms

© Springer Nature Switzerland AG 2020   65
C. Carlberg et al., *Nutrigenomics: How Science Works*,
https://doi.org/10.1007/978-3-030-36948-4_5

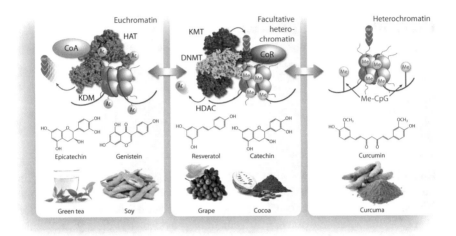

Fig. 5.1 Chromatin, associated proteins and nutrition-based epigenetic modulators.
Chromatin is distinguished into open chromatin (euchromatin, **top left**) with loose nucleosome
arrangement and closed chromatin represented by dense nucleosome packing (heterochromatin,
top right). There are several stages between these extremes that are summarized as facultative
heterochromatin (**top center**). Each stage is characterized by a set of chromatin modifiers, such as
HATs, HDACs, HMTs, HDM (histone demethylases) and DNMTs (DNA methyltransferases), that
lead together with CoA (co-activator) and CoR (co-repressor) proteins to the schematically indicated scenarios of acetylation and methylation of histone tails and genomic DNA. The indicated
plant-origin natural compounds have been shown to modulate the activity of chromatin modifiers
(**bottom**) and to affect in this way the epigenetic status of cells and tissues

are post-translational modifications of nucleosome-forming histone proteins (Box
5.1) and methylation of cytosines within genomic DNA (Box 5.2).

Chromatin density plays an important role in regulating gene expression,
primarily by controlling the accessibility of genomic binding sites for transcription
factors. The dynamic competition between nucleosomes and transcription factors
for critical binding sites is influenced by a large set of enzymes that either cova-
lently modify histone proteins, termed chromatin modifiers (Box 5.3), or move,
reconfigure or eject nucleosomes, called chromatin remodelers. This process determines for each genomic region the density, composition and positioning of nucleosomes relative to the transcription factor binding sites that it contains.

The *ENCODE Project* and other big biology consortia (Box 2.4) provide
genome-wide maps of epigenetic marks in more than 100 human cell lines, primary
cells and tissues. These data collectively indicate that in regions of open, active
chromatin histone proteins are acetylated (*e.g.*, at lysine 14 of histone 3, H3K14ac)
and genomic DNA remains unmethylated. In contrast, in repressed, closed chromatin histones are (tri)methylated (*e.g.*, at lysine 27 of histone 3, H3K27me3) and also
the DNA gets methylated. The change in DNA methylation during development
starts with demethylation during cell divisions of the fertilized egg, followed by *de
novo* methylation after implantation. **Due to this epigenetic reprogramming**

Box 5.1: Histone Proteins and Their Modifications

The basic, every 200 bp repeating unit of chromatin is the nucleosome, which consists of 147 bp genomic DNA that is wrapped nearly twice around an octamer of four pairs of the histone proteins H2A, H2B, H3 and H4 (Fig. 5.1). Histone proteins are rich in lysine residues, in particular at their amino-terminal tails that stick out from the nucleosome. The lysine residues but also arginine and serine residues are subject to post-translational, covalent, reversible chemical modifications, such as acetylations, methylations and phosphorylations, carried out by a variety of chromatin modifiers (Box 5.3). These histone modifications, referred to as epigenetic marks, represent a kind of chromatin indexing. Most marks are assigned to functional regions of the chromatin, such as promoters, enhancers or heterochromatin. The rules for this chromatin indexing are summarized in the histone code, which is the major determinant for the accessibility of genomic binding sites of transcription factors and their associated co-factors. The final outcome is increased or decreased gene expression, *i.e.*, mRNA levels. Many chromatin modifiers and other nuclear proteins contain a set of common domains that specifically recognize different chromatin modifications, *i.e.*, these proteins are able to "read" the histone code (Sect. 5.2).

Box 5.2: DNA Methylation

From the four DNA forming nucleotides only cytosines gets methylated and this in particular at the dinucleotide CpG, *i.e.*, at sites where cytosine and guanine are found on the same DNA strand and are connected by a phosphodiester bond. CpG islands are genomic regions of at least 200 bp in length displaying a CG content of at least 55% (compared to the 42% average in the human genome). In normal human cells, CpG islands are mostly unmethylated and the genes in their vicinity keep their potential to be activated by transcription factors. DNMTs use SAM (S-adenosylmethionine) as a methyl group donor (Sect. 5.3) to methylate the carbon in 5'-position of cytosines. DNMT1 is a maintenance methyltransferase being active mainly during DNA replication, while DNMT3A and DNMT3B primarily perform *de novo* DNA methylation during embryogenesis and cellular differentiation. The demethylation of genomic DNA is regulated by TET (ten-eleven translocation) proteins, which convert 5-methylcytosine to 5-hydroxymethylcytosine, 5-formylcytosine and 5-carboxylcytosine. During DNA replication these modifications are then removed through multi-step oxidation and base excision repair.

Box 5.3: Chromatin Modifying Enzymes
The activity of chromatin is modulated by a group of enzymes that catalyze rather minor changes in residues of histone proteins, such as the addition or removal of acetyl or methyl groups. Chromatin acetylation is generally associated with transcriptional activation, while the exact amino acid residue of the histone tails that is acetylated is not critical. The acetylation state of a given chromatin locus is controlled by two classes of antagonizing histone modifying enzymes, HATs and HDACs. In analogy, also for histone methylation there are two classes of enzymes with opposite functions, HMTs and HDMs. Although histone methylation mainly mediates chromatin repression, at certain residues, such as H3K4, it results in activation. Therefore, for histone methylation the exact residue in the histone tail and its degree of methylation (mono-, di- or tri-methylation) is of critical importance.

during development, the intrauterine period is considered critical for long-term health and disease risk (Sect. 5.4). However, DNA methylation does not only control the expression of specific genes during the development and differentiation of individual tissues, but it is also essential for silencing of imprinted genes, the second female X chromosome and retrotransposons (Box 2.3). Importantly, histone modifications precede or succeed DNA (de)methylations, *i.e.*, both epigenetic processes display a cross-talk. For example, methyl-CpG binding proteins are capable of recruiting HDACs to methylated regions of genomic DNA.

5.2   Intermediary Metabolism and Epigenetic Signaling

Within a cell, signal transduction pathways result in the activation of gene expression programs that integrate signals originating from the environment. This could be the availability of energy substrates, which induce responses of a larger set of genes, in order to maintain homeostasis of the organism. Numerous connections between products of intermediary metabolism and chromatin modifiers are known. Our genome expresses tissue-specifically more than hundred of these enzymes that interpret ("read"), add ("write") or remove ("erase") post-translational histone modifications (Fig. 5.2). The activity of most of these chromatin modifiers critically depends on intracellular levels of essential metabolites, such as acetyl-CoA, UDP (uridine diphosphate)-glucose, α-ketoglutarate, NAD^+, FAD (flavin adenine dinucleotide), ATP or SAM. Since the cellular concentrations of several of these metabolites represent the metabolic status of the cell, **the activities of the chromatin modifiers reflect the intermediary metabolism.**

   A wide spectrum of secondary metabolites from fruits, vegetables, teas, spices, and traditional medicinal herbs, such as genistein, resveratrol, curcumin and polyphenols from green tea, coffee and cocoa, respectively, are able

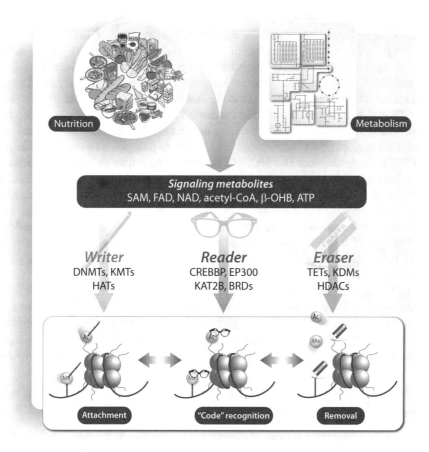

Fig. 5.2 Epigenetic mechanisms link metabolites and transcription. Changes in nutrition or fluctuations in metabolism affect the transcriptional responses of metabolic tissues. Several intermediary metabolites change the activity of chromatin modifiers in a dose-dependent manner. These proteins use some of these metabolites as co-substrates and/or co-factors and act in this way as metabolic sensors. "Writer" enzymes create covalent chromatin marks, "reader" enzymes recognize these marks and "eraser" enzymes remove them. These histone tail modifications create changes in the local chromatin structure, which has consequences for the activity and regulation of the neighboring genes

to modulate the activity of transcription factors and chromatin modifiers (Fig. 5.1, bottom). Next-generation sequencing technologies allow the genome- and transcriptome-wide assessment of the specificity and efficacy of these compounds, *e.g.*, for preventing and/or treating cancer.

Bromodomains are found in all type of proteins that are able to recognize acetylated residues, such as members of the BRD (bromodomain containing) protein family, the HATs KAT2B, EP300 (KAT3B) and CREBBP (CREB binding protein, also called KAT3A), HMTs, chromatin remodeling enzymes, CoAs and EP300, and general transcription factors. In contrast, chromodomains are far more specific for a given chromatin modification, *i.e.*, chromodomain-containing nuclear proteins recognize their genomic targets with far more accuracy than bromodomain proteins.

Methyl donors are critical during pregnancy and dietary excess as well as deficiency may have an impact on epigenetic programming in mice (Sect. 5.3) and humans (Sect. 5.4). The methyl donor substrate SAM connects DNA methylation with intermediary metabolism. SAM is generated from the amino acid methionine and ATP in the one-carbon metabolism pathway (Box 5.4). When a methyl group of SAM is transferred to DNA or a histone, the product SAH (S-adenosylhomocysteine) is recycled back to SAM. Interestingly, SAH acts as a negative feedback regulator for HMTs, suggesting that **the SAM/SAH ratio, referred to as the "methylation index", is critical for histone and DNA methylation**. A derivative of the B vitamin folate, tetrahydrofolate, serves as a methyl group donor for feeding of the cyclic one-carbon pathway. The dependence of the pathway on folate and other micronutrients is another example of the direct connection between nutrition and epigenetics. The function of folate in normal neural tube closure in early gestation (in human 21–28 days after conception) is well known and maternal supplementation with folate is recommended for prevention of neural tube defects.

The energy status of our tissues and cell types is the most important information for our body, in order to interpret and integrate environmental conditions. Nutritional epigenetics investigates how metabolic pathways communicate with chromatin and provide information about nutrient availability and energy status. Since key metabolites, such as AMP, NAD^+, SAM and acetyl-CoA, act as

---

Box 5.4: Folate Metabolism and Methylation

A derivative of the B vitamin folate, tetrahydrofolate, feeds the cyclic one-carbon pathway by serving as a methyl group donor. This demonstrates a direct connection between nutrition and epigenetics. **Methyl group donors are critical for epigenetic programing during embryogenesis**. A high homocysteine level is an established biomarker for the disturbance of the one-carbon metabolism and related to low concentrations of folate, vitamins B6 and B12, choline and betaine. This may cause an elevated risk of premature delivery, low birth weight and neural tube defects. Moreover, a low dietary intake of folate or methionine increases the risk of colon adenomas, while *in utero* exposure to higher folate is associated with a reduced risk of childhood acute lymphoblastic leukemia, brain tumors and neuroblastoma. The enzyme methylenetetrahydrofolate reductase, which is encoded by the *MTHFR* gene, catalyzes the conversion of 5,10-methylenetetrahydrofolate to 5-methyltetrahydrofolate. 10–15% of Europeans carry the missense SNP rs1801133 on both alleles, which reduces the activity of the enzyme by more than 50%. Accordingly, individuals with a TT genotype are affected more by a low folate intake than those with the CC or CT allele. Accordingly, raised plasma homocysteine concentrations cause an elevated risk of premature delivery, low birth weight and neural tube defects.

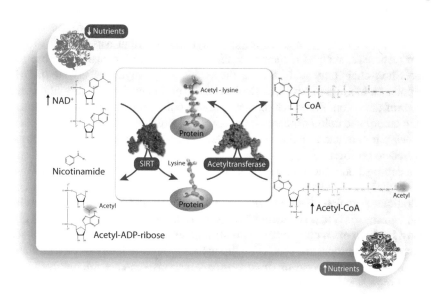

Fig. 5.3 The relation of protein acetylation and cellular metabolism. NAD⁺ (**left**) acts as a
co-factor for HDACs of the sirtuin family that deacetylate proteins, which had been acetylated by
HATs using acetyl-CoA (**right**). Thus, the acetylation status of key regulatory proteins reflects the
cellular concentration of NAD⁺ and acetyl-CoA, *i.e.*, of low (**top**) or high (**bottom**) nutritional
status, respectively

co-factors and substrates of chromatin modifiers, gene expression programs of
many central physiological processes, such as proliferation and differentiation, are
modulated by the metabolic status of the cells. The results of these epigenetic
events may be memorized in the epigenome of skeletal muscle and adipose tissue.
The latter two metabolic organs constitute more than half of our body mass.
However, their relative amount is very variable and depends on environmental factors, such as physical activity and nutritional intake. This means that **our lifestyle
creates a metabolic memory**. Thus, not only the tissue mass but also **the epigenome of our muscles and fat memorizes how much we have eaten and
exercised**.

The ratio of the oxidized (NAD⁺) and reduced (NADH) form of the co-factor
NAD reflects the cellular redox state and is inversely proportional to the energy state
of a cell. During fasting, *i.e.*, at low levels of nutritional metabolites, the intracellular concentration of NAD⁺ raises. This leads to an increase in the activity of HDACs
of the SIRT family (which use NAD⁺ as a co-factor) and the deacetylation of their
target proteins (Fig. 5.3, left). The targets are often histones, but also transcription
factors or their co-factors, such as p53 and PPARGC1A, are affected in their acetylation status. **Calorie restriction, *i.e.*, using only 70–80% of the recommended
dietary intake, is beneficial for metabolic health and may slow down aging**

(Sect. 6.5). Since the NAD$^+$ concentrations fluctuate in a circadian manner, sirtuin-mediated gene regulation is linked to the epigenetic clock (Sect. 3.6). Accordingly, time-restricted feeding can restore daily rhythms and improves number of metabolic responses, such as reducing insulin resistance and increasing glucose tolerance. Short-chain fatty acids, such as the ketone body β-hydroxybutyrate (β-OHB), are potent inhibitors of several HDACs and are produced from dietary fibers in the lumen of the colon. Fiber-rich diet can prevent colitis (*i.e.*, a chronic inflammation of the colon) and colon cancer in human, which could be, at least in part, explained *via* butyrate-mediated inhibition of HDAC-dependent transcriptional programs of colonocyte proliferation.

In contrast, nutrients ingested in the feeding state enter the catabolic pathways of intermediary metabolism and acetyl-CoA is produced. Augmented acetyl-CoA concentrations stimulate HAT activity, so that their target proteins get acetylated (Fig. 5.3, right). **When the target proteins are histones, the acetylation of chromatin leads to open chromatin**. This stimulates the expression of genes involved in metabolic processes, such as lipogenesis and adipocyte differentiation. Moreover, a large proportion of acetyl-CoA-responsive genes are involved in cell cycle progression, *i.e.*, an increase in histone acetylation is associated with cellular proliferation. However, upon induction of cellular differentiation the acetyl-CoA level decreases significantly. Accordingly, loss of pluripotency is associated with decreased glycolysis and lower levels of acetyl-CoA and histone deacetylation. Moreover, the acetyl-CoA level also affects cell survival and death decisions. For example, a low acetyl-CoA level induces the catabolic process of autophagy (Box 3.1), which is crucial for organelle quality control and cell survival during metabolic stress. Thus, **the acetyl-CoA/CoA ratio is an important regulator of major cellular decisions**.

Another example of metabolite sensing is that of the enzyme AMPK, which is controlled in its activity by the AMP/ATP ratio (Sect. 6.6). When cells consume more ATP than they are producing, *i.e.*, at conditions of low nutrient availability, AMP concentrations raise as a signal of energetic stress. AMP binds to the γ-subunit of the AMPK heterotrimer and activates the kinase. Since histones are AMPK substrates, a low energy status of the cell is marked *via* histone phosphorylation. Thus, **insults to the energy status of a cell are memorized on the level of histone modifications and can be translated into functional outputs *via* adaptive gene regulation**. In contrast, a high nutritional level results in low AMP levels, no AMPK activity, a modified histone phosphorylation pattern and the activity of a different set of genes. Thus, the metabolic state of a cell can be expressed by the ATP/AMP ratio, the SAM/SAH ratio, the NADH/NAD$^+$ ratio and the acetyl-CoA/CoA ratio (Fig. 5.4). Under high nutrient concentrations, such as abundant availability of methionine and glucose, SAM activates KMTs and acetyl-CoA stimulates HATs, thus leading to histone methylation and acetylation, respectively. In contrast, at low nutrient levels, such as during fasting, AMP activates AMPK and NAD$^+$ stimulates sirtuins resulting in histone phosphorylation and deacetylation. Moreover, in parallel SAH inhibits DNMTs and CoA blocks HATs.

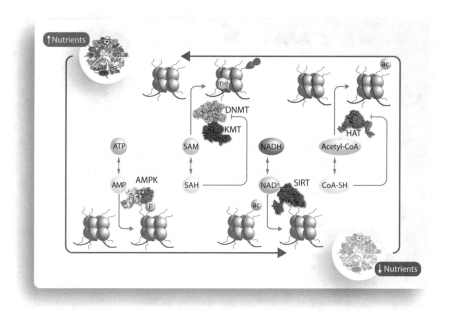

Fig. 5.4 Epigenetic sensing of the nutritional state. A high nutritional state of a cell (**top**) is represented by the abundance of the metabolites ATP, SAM, NADH and acetyl-CoA, while in the case of low nutrient levels (**bottom**) the metabolites AMP, SAH, NAD$^+$ and CoA are predominant. Accordingly, at high nutrient concentrations, KMTs and HATs are stimulated, while at low concentrations AMPK and HDACs of the sirtuin family are activated and DNMTs and HATs are repressed. This results in histone methylation and acetylation or histone phosphorylation and deacetylation, respectively

5.3   Nutrition-Triggered Transgenerational Epigenetic Inheritance

The epigenome is able to preserve the results of cellular perturbations by environmental factors in form of changes in DNA methylation, histone modifications and 3D organization of chromatin. Thus, **the epigenome has memory functions.** Changes in epigenomic patterns, such as DNA methylation maps, are called epigenetic drifts (Sect. 5.4). They primarily describe the lifelong information recording ("experience") of somatic cell types and tissues, but may also be inherited to daughter cells, when the cells are proliferating. In case of germ cells, epigenetic drifts may be, at least in part, even transferred to the next generation. This leads to the concept of transgenerational epigenetic inheritance, which suggests that the lifestyle of the parent and grandparent generation, such as daily habits in food intake or physical activity, may affect their offspring.

The master example of the transgenerational epigenetic inheritance concept is the agouti mouse model. The *Asip* gene encodes for peptide hormone that stimulates

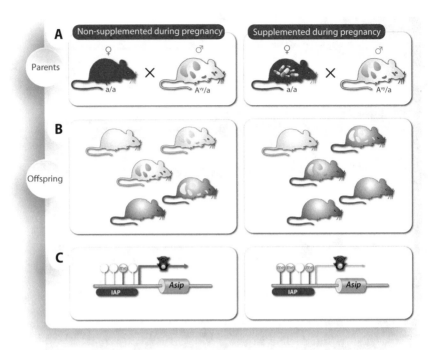

Fig. 5.5 Maternal dietary supplementation affects the phenotype and epigenome of A^vy^/a offspring. The diets of female wild-type a/a mice are either not supplemented (**left**) or supplemented with methyl-donating compounds (**right**), such as folate, choline, vitamin B12 and betaine, for 2 weeks before mating with male A^vy^/a mice, and ongoing during pregnancy and lactation (**a**). The fur color of offspring that are born to non-supplemented mothers is predominantly yellow, whereas it is mainly brown in the offspring from mothers that were supplemented with methyl-donating substances (**b**). About half of the offspring does not contain an A^vy^ allele and is therefore black (a/a, not shown here). Molecular explanation of DNA methylation and *Asip* gene expression: maternal hypermethylation after dietary supplementation shifts the average fur color distribution of the offspring to brown by causing the IAP retrotransposon upstream of the *Asip* gene to be more methylated on average than in offspring that are born to mothers fed a non-supplemented diet (**c**). White circles indicate unmethylated CpGs and yellow circles are methylated CpGs

in melanocytes close to hair follicle the synthesis of yellow pheomelanin instead of black or brown eumelanin, *i.e.*, the fur of the respective mice gets yellowish. Moreover, the ASIP protein is involved in the neuronal coordination of appetite acting as the physiological antagonist of MC1R (Sect. 4.1). In the transgenic agouti mouse model the retrotransposon IAP (intracisternal A particle) was inserted into the regulatory region of the *Asip* gene (Fig. 5.5). This creates a dominant allele of the *Asip* gene (termed A^vy^), the expression of which is depending on the methylation level of IAP, *i.e.*, on its epigenetic status. The methylation of the IAP retrotransposon happens stochastically during early embryogenesis, *i.e.*, A^vy^ behaves as a metastable epiallele. Heterozygous A^vy^/a mice vary in their fur colors from yellow via mottled to wild-type dark fur color. When the IAP retrotransposon is methylated, the synthesis of pheomelanin is downregulated and a dark fur color appears. In contrast, non-methylated IAP allows ubiquitous *Asip* gene expression leading to both yellow fur color and obesity. When A^{vy}/a mice inherit the A^{vy} allele maternally, *Asip* gene expression and fur color correlate with the maternal phenotype. Thus, the fur color provides an easy phenotypic readout of the epigenetic status of IAP throughout life. This makes the A^{vy} *Asip* mouse an ideal *in vivo* model for the investigation of a mechanistic link between environmental stimuli, such as nutrition, and epigenetic states of the genome.

The agouti mouse model was used for the following experiment: 2 weeks before mating with male A^{vy}/a mice, female wild-type a/a mice were either supplemented or not with methyl donors, such as folate, vitamin B12 and betaine (Fig. 5.5, Box 5.4). The supplementation was continued during pregnancy and lactation. While the F1 generation of non-supplemented mothers displayed the expected number of yellow color phenotypes, the offspring of supplemented mothers shifted toward a brown fur color phenotype. This suggests that maternal methyl donor supplementation leads to increased A^{vy} methylation in the offspring. Furthermore, this means that an environmentally induced epigenetic drift in the mothers was inherited to their children. The inheritance of an epigenetic programming to the next generation indicates that at least metastable epialleles, such as IAP, are able to resist the global demethylation of the genome before preimplantation. Taken together, the model indicates that **epigenetic memory can be passed from one generation to another by inheriting the same indexing of chromatin marks**. From the different types of chromatin marks, DNA methylation is designed in particular for a long-term cell memory, while short-term "day-to-day" responses of the epigenome are primarily mediated by non-inherited changes in the histone acetylation level. Histone methylation levels are in between both extremes.

The mouse models impose the question whether the concept of an epigenetic memory and inheritance is also valid for humans. There are no comparable natural human mutants and human embryonal feeding experiments cannot be performed for ethical reasons. However, there are natural "experiments", such as the *Dutch Hunger Winter*, where individuals were exposed *in utero* to an extreme undernutrition occurring in the Netherlands during the winter of 1944/45. Fetal malnutrition led to impaired fetal growth. Low birth weight favors a thrifty phenotype that is epigenetically programmed to use nutritional energy efficiently, *i.e.*, to be prepared for a future environment with low resources during adult life. Even many decades after birth the *in utero* undernourished individuals showed subtle (<10%) changes in DNA methylation at several loci in adulthood, for example, at the regulatory region of the imprinted gene insulin-like growth factor 2 (*IGF2*). This epigenetic pattern is associated with an increased risk of obesity, dyslipidemia and insulin resistance, when the respective individuals are exposed to an obesogenic environment. Accordingly, **for humans there is the same link between prenatal nutrition and epigenetic changes as described for rodents**. Similarly, the famine of 1959–1961 in China largely contributed to the overproportional high raise of T2D in the country. These examples led to the DOHaD (Developmental Origins of Health and Disease)

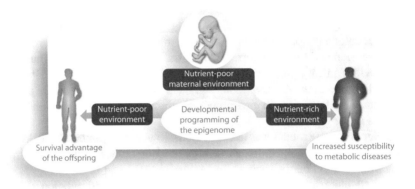

Fig. 5.6 The DOHaD concept. Intrauterine stressors, including maternal undernutrition or placental dysfunction (leading to impaired blood flow with consecutively hypoxia or reduced nutrient transport) can initiate abnormal patterns of development, histone modifications and DNA methylation. Additional post-natal environmental factors, including accelerated post-natal growth, obesity, inactivity and aging further contribute to the risk for T2D, potentially via changes in histone modifications and DNA methylation patterns of metabolic tissues. Obviously, epigenetic changes during embryogenesis have a much greater impact on the overall epigenetic status of an individual than that of adult stem cells or somatic cells, since they affect far more following cell divisions

concept (Fig. 5.6) indicating that early developmental events, such as perturbations of the nutritional state *in utero*, have significant effects on disease risk as adult. Thus, **environmental exposures of individuals, in particular during early life, can be stored as epigenetic memory**. In particular, when post-natal environment, such as being obesogenic, differs from what the epigenome was programmed in the prenatal phase, such as starvation, responses of the body may be maladaptive, like developing obesity and T2D.

5.4 Population Epigenetics

The field of epigenetic epidemiology, *i.e.*, the study of epigenetics in populations, combines epigenome-wide methods (Sect. 2.5) with population-based epidemiological approaches. Epigenetic changes can occur at any time during life, although increased sensitivity exist during early embryogenesis. **Our epigenome primarily changes due to environmental exposures but also based on stochastic epigenetic drifts associated with aging** (Sect. 6.5). Epigenetic epidemiology studies different types of human cohorts with the goal to identify both the causes as well as the phenotypic consequences of epigenomic variations. These are

- cohorts of natural experiments, such the *Dutch Hunger Winter*
- longitudinal birth cohorts
- longitudinal studies on monozygotic twin cohorts
- prenatal cohorts
- *IVF (in vitro* fertilization) conception cohorts.

Studies of families are well suited to investigate epigenetic changes in the offspring that may be based on environmental exposures of parents during gametogenesis. Birth cohorts and *in vitro* fertilization cohorts track life from as early as periconception (*i.e.*, around the time of conception) and allow the study of epigenetic changes based on the prenatal environment and their association with disease phenotypes early in life. Cohorts based on natural experiments, *i.e.*, when the exposure to severe environmental conditions was not under experimental control, enable the investigation of the link of environmental exposures in early life with the onset of disease phenotypes decades later. Prospective cohorts, in particular those involving monozygotic twins having identical genomes, parents, birth date and gender, study in a longitudinally way the contribution of age-related epigenetic modifications in common diseases, such as T2D, autoimmune diseases and cancer. Interestingly, twin studies demonstrated that epigenetic variations significantly increase across lifespan. Short-term interventions, such as dietary studies, can identify specific environmental exposures that lead to tissue-specific epigenetic changes (Fig. 5.7).

Epigenetic drifts, such as hypermethylation of CpG islands close to the regulatory regions of tumor suppressor genes, contribute to the risk for cancer and other diseases (Fig. 5.8). In particular the risk for diseases that are related to the exposure with environmental factors, such as microbes causing inflammation or overeating leading to obesity and T2D, have a large epigenetic contribution. Of special interest are diseases that have their onset a long time before the phenotype emerges, *i.e.*, where accumulation of epigenetic changes stepwise increase disease susceptibility, such as Alzheimer's disease and cancer.

Epigenetic information can also be transmitted *via* the germ line to the next generation. Although some 95% of DNA methylation marks are erased throughout the two rounds of demethylation during primordial germ cell generation, some single-copy genomic regions escape both demethylation waves and remain methylated in gametes. In case the DNA methylation pattern at these escape regions is susceptible to environmental influences, such as dietary molecules, **lifestyle choices of an individual may be transmitted to subsequent generations and may lead to phenotypic consequences**.

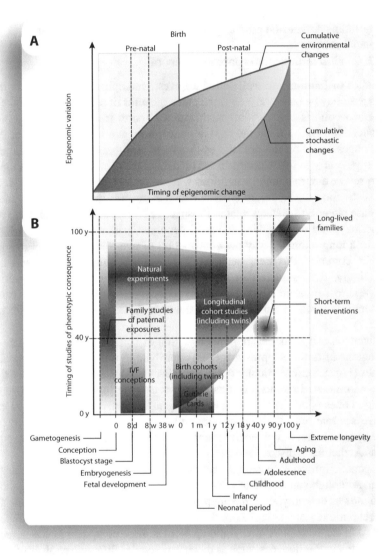

Fig. 5.7 Epigenetic variation in populations. Epigenetic changes can occur at any time during life, but there is significantly increased sensitivity during early prenatal development (**a**). Prenatal epigenetic changes may be investigated using IVF cohorts, archived newborn metabolic screening tests (formerly called Guthrie cards) and birth cohorts tracking life from as early as periconception (**b**). Historical famines represent the few opportunities to link the prenatal environment to health outcomes later in life. Longitudinal cohort studies (especially involving twins) sample peripheral tissues and take biopsies from disease-relevant tissues. Short-term (dietary) interventions can iden-tify specific dietary compounds that induce tissue-specific epigenetic modifications, while long-lived families can help in identifying the importance of maintaining epigenetic control for healthy aging

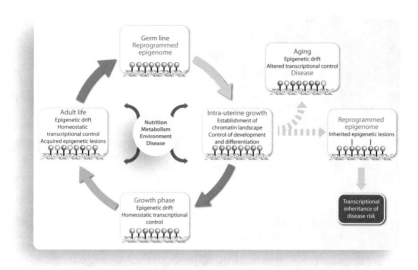

Fig. 5.8 Epigenetic drift and transgenerational inheritance. During embryogenesis epigenetic marks, such as DNA methylation and histone modifications, are established in order to maintain cell lineage commitment. After birth, this epigenetic landscape stays dynamic throughout lifespan and responds to nutritional, metabolic, environmental and noxious signals. Epigenetic drifts are part of homeostatic adaptations and should keep the individual in good health. However, when an adverse epigenetic drift compromises the capacity of metabolic organs to adequately respond to challenges as provided by nutrition and inflammation, the susceptibility to diseases, such as T2D or cancer, increases. Some of these acquired epigenetic marks can be inherited to subsequent generations when they escape epigenetic reprogramming during gametogenesis

Additional Readings

Barres R, Zierath JR (2016) The role of diet and exercise in the transgenerational epigenetic landscape of T2DM. Nat Rev Endocrinol 12:441–451

Carlberg, C., and Molnár, F. (2018). Human epigenomics. Springer Textbook Springer. ISBN: 978-981-10-7614-8

Gut P, Verdin E (2013) The nexus of chromatin regulation and intermediary metabolism. Nature 502:489–498

Heard E, Martienssen RA (2014) Transgenerational epigenetic inheritance: myths and mechanisms. Cell 157:95–109

Kinnaird A, Zhao S, Wellen KE, Michelakis ED (2016) Metabolic control of epigenetics in cancer. Nat Rev Cancer 16:694–707

Sales VM, Ferguson-Smith AC, Patti ME (2017) Epigenetic mechanisms of transmission of metabolic disease across generations. Cell Metab 25:559–571

Chapter 6
Nutritional Signaling and Aging

Abstract   In this chapter, we will present the evolutionary conservation of nutrition-sensing pathways and their relation to the process of aging. Mammals use more complex regulatory circuits for sensing food, which involve the CNS *via* the GH1 endocrine axis. The molecular basis of this is the sensing of glucose and amino acids *via* insulin/IGF and the TOR pathways, respectively, and the integration of the nutritional and energetic status of cells and tissues *via* sirtuins and AMPK. The insulin signaling axis is composed of a number of critical nodes including the receptor IR, the adaptor protein IRS, the kinases PI3K and AKT as well as the transcription factor FOXO1. We will analyze the mechanisms how this central signal transduction pathway interacts with environmental challenges mediated *via* multiple other pathways, in order to keep cells and tissues in homeostasis. Under conditions of calorie restriction, *i.e.*, at reduced food intake, the lifespan of model organisms, such as yeast, worms or flies, is increased. Interestingly, signal transduction pathways related to calorie restriction show also in humans very similar regulatory principles. **This insight has the potential to prevent age-related diseases, such as T2D, CVDs and cancer, and to promote healthy aging in human**.

Keywords Aging · Model organisms · Nutrient sensing · Insulin/IGF signaling · TOR-S6K signaling · Growth hormone endocrine axis · IR · IRS · PI3K · AKT · FOXO1 · Calorie restriction · Sirtuins · NAD · AMPK · Cellular energy status

6.1   Aging and Conserved Nutrient-Sensing Pathways

Aging is a complex molecular process that affects all species. It is represented by the accumulation of molecular, cellular and organ damage, leading to loss of function and increased risk to disease and finally death. Nutrient-sensing pathways are fundamental to the aging process. Abundance of food activates nutrient-sensing pathways that stimulate a diverse set of physiological processes. The latter includes reproduction but is compromised by a limited lifespan. In contrast, in conditions of starvation or when the nutrient-sensing pathways are genetically interrupted, reproduction is delayed and the lifespan increased. Thus, **the availability of food determines the speed of aging**.

Understanding the molecular basis of aging is a central topic of nutrigenomics. Nevertheless, aging research in humans takes time and ethical reasons restrict many types of experiments. Therefore, most of the principles of aging were first understood *via* the use of simple model organisms, such as *Saccharomyces cerevisiae* (yeast), *Caenorhabditis elegans* (roundworm), *Drosophila melanogaster* (fruit fly) and *Mus musculus* (mouse) that have a far shorter lifespan than humans (Box 6.1).

Box 6.1: Model Organisms

A model organism is a non-human species that is studied *in vivo*, in order to understand biological processes, such as aging, that due length of lifespan, costs or ethical reasons cannot be studied in humans. The evolutionary conservation of biological pathways allows the transfer of at least some of the results and insights obtained with the model organisms to humans. In the unicellular species yeast, not only the survival of a population of non-dividing cells (chronological lifespan) can be studied, but also the number of daughter cells generated by a single mother cell (replicative lifespan). The roundworm *C. elegans* is a simple multicellular species formed only by some 1000 cells but already allows studies of different cell types and organs, such as nervous or digestive systems. *C. elegans* was the first species, in which lifespan extending mutations were found. The fruit fly *D. melanogaster* confirms the evolutionary conservation of biological pathways. It contains more different tissues than *C. elegans* and allows the examination of sex differences. Finally, the mouse is the most established model organism in biomedical research and as a mammalian species it is much closer to humans (even though primates are closest, but for ethical and generation length reasons, research with them is limited).

Multicellular organisms have developed a nutrient-sensing system allowing communication between different parts of the body. The intracellular signal transduction pathway of the peptide hormone IGF1 is the same as that of insulin, and informs cells about the presence of glucose (Sect. 9.1). The insulin/IGF pathway (Sect. 6.3) is evolutionary very conserved and, depending on the respective species, starts with one or several specific receptor tyrosine kinase-type membrane receptors (Fig. 6.1). *Via* cytosolic adaptor proteins, such as IRS (insulin receptor substrate), and kinases, such as PI3K (phosphoinositide 3-kinase) and AKT (AKT serine/threonine kinase), the ligand stimulation of the receptors results in the inactivation of one or several members of the FOXO transcription factor family. FOXOs that control the expression of genes involved in a wide range of physiological processes, such as cellular stress response, anti-microbial activity and detoxification of xenobiotics and free radicals (Sect. 6.4). At least in lower organisms, such as *C. elegans*

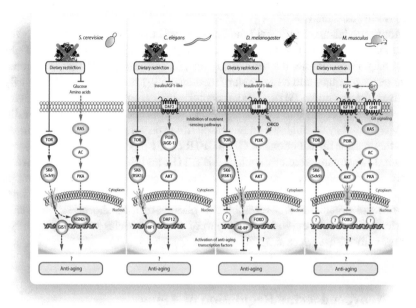

Fig. 6.1 Nutrient signaling pathways involved in longevity are conserved in various species.
The activity of various signal transduction pathways is reduced by calorie restriction either directly
(in yeast) or indirectly (in worms, fruit flies and mice) through the reduced levels of growth factors,
such as IGF1. In all four species TOR and S6K activation promote aging (*i.e.*, reduce lifespan).
Moreover, in yeast and mammals activation the AC (adenylate cyclase)-PKA (protein kinase A)
pathway accelerates aging. In addition, aging is promoted by the insulin/IGF1 signaling pathway
directly or indirectly *via* its upstream factors, such as GH1. Transcription factors, such as GIS1
(GIg1–2 suppressor) and MSN (multicopy suppressor of SNF1 mutation) 2/4 in yeast, DAF
(abnormal dauer formation) 12 and HIF1 (hypoxia-inducible factor 1) in worms and FOXO in flies
and mice as well as the translational inhibitor 4E-BP in flies are inactivated by either the AC-PKA,
the IGF1-AKT or the TOR-S6K pathway. They protect from aging in all major model organisms

and *D. melanogaster*, the **knockout of any gene/protein that leads to the specific
interruption of this signal transduction pathway causes lifespan extension**.

Parallel to the glucose-sensing system there is an amino acid-sensing pathway
that is also evolutionary highly conserved (Fig. 6.1). Central to this signal
transduction pathway are the proteins TOR and S6K (S6 kinase). TOR is a cell-
growth modulator and serine/threonine kinase, which in mammalians exists in two
different complexes, TORC1 and TORC2, of which only TORC1 is sensitive to
amino acids (Sect. 3.1). The TOR-S6K pathway demonstrates crosstalk with the
insulin/IGF pathway and the inhibition of its activity also can increase lifespan, at
least in lower organisms. In mice, mutations in genes for GH1 and the insulin/IGF
signaling pathway can increase lifespan by up to 50%. Moreover, the inhibition of
the TOR pathway increases the lifespan of mice and at the same time reduces the

incidence of age-related pathologies, such as bone, immune and motor dysfunctions and insulin resistance.

Mouse models have indicated that in mammals the sensing of nutrition involves additional control circuits, including actions of the CNS and its associated glands (Sect. 6.2). For example, the somatotrophic axis comprises GH1 that is secreted by the anterior pituitary and its secondary mediator IGF1 that is produced primarily by the liver. Interestingly, GHR (GH1 receptor)-deficient primates develop seldomly diabetes or cancer but due to developmental defects and the increased mortality at younger ages they do not have an increased life expectancy. However, genetic variations that reduce the functions of GH, IGF1R (IGF1 receptor), IR (insulin receptor) or their downstream effectors, such as AKT, TOR and FOXO1 (Fig. 6.1), have been linked also to longevity in humans. The study of conserved nutrient-sensing pathways suggests that the genetic alterations create a physiological state in the investigated model organisms that resembles to periods of food shortage. Thus, **calorie restriction is able to extend the lifespan of diverse species spanning from yeast to rhesus monkeys** (Sect. 6.5).

6.2   Neuroendocrine Regulation of Aging

Aging starts for humans after the peak of their maximal physical ability in the age of 20–25 years, *i.e.*, already at this age begins a slow progressive decline of the physiological capabilities of the different tissues and cell types forming our body. From an age of 45 years onwards, there is a significant increase in the onset of aging-related complex diseases, such as cancer, T2D, CVDs and Alzheimer's disease. For many thousand years the average generation length of *Homo sapiens* was approximately 20 years and increased just in the recent past. Assuming some additional 25 years for raising all offspring, **only within the first 45 years of life evolutionary adaption had the chance to select for sufficient fitness and health**. This means that any harm occurring to humans above this age, such as developing T2D or cardiovascular problems, can not be corrected *via* evolutionary adaption principles, such as an increased or decreased number of vital offspring (Sect. 2.2). Nevertheless, the maximal lifespan of humans extends to approximately 120 years. The extra time of 75 years, which was realized only in a few subjects, can be considered as a margin of safety, in order to guarantee that the vast majority of humans have enough time to fulfill the primary evolutionary sense of life, *i.e.*, **to reproduce and to assure the survival of their children until these start reproduction themselves**.

GH1 is vertebrate-specific peptide hormone that is produced in the pituitary gland. Loss-of-function mutations in genes encoding for transcription factors important for pituitary development, such as POU1F1 (POU class 1 homeobox 1) and PROP1 (PROP paired-like homeobox 1), result in deficiency in *GH1* expression and large lifespan extension, which is mediated, at least in part, by insulin/IGF signaling. Thus, **a defensive response of minimized cell growth and metabolism in**

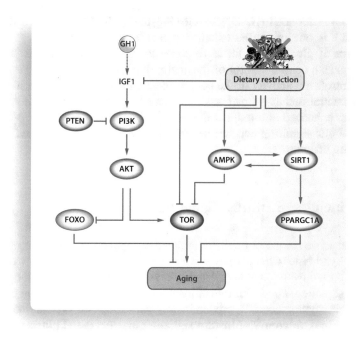

Fig. 6.2 Nutrient-sensing in mammals. Overview of the endocrine somatotrophic axis that involves GH1 and the insulin/IGF1 signal transduction pathway. Proteins/pathways that favor aging are shown in green and those with anti-aging properties in red

case of cellular damage or food shortage enables an organism with a constitutively decreased insulin/IGF signaling to survive longer.

For the same reason physiologically or pathologically aged mammals decrease their insulin/IGF signaling, *i.e.*, during normal aging the levels of GH1 and IGF1 decline. In addition to insulin/IGF signaling sensing glucose, also in mammals high amino acid concentrations are sensed by TOR and low-energy states by sirtuins *via* high NAD⁺ levels and by AMPK *via* high AMP levels (Sect. 6.6) (Fig. 6.2). During aging, TOR activity increases in hypothalamic neurons and contributes to age-related obesity, which in mice can be reversed by infusion of the TOR inhibitor rapamycin to the hypothalamus. However, TOR inhibition creates side effects, such as impaired wound healing and insulin resistance, *i.e.*, this pathway is not suited for pharmacological intervention. In contrast, sirtuins and AMPK are counteracting to insulin/IGF and TOR signaling, *i.e.*, their activity represents low food availability and catabolism instead of nutrient abundance and anabolism. Interestingly, one of the activities of AMPK is the direct inhibition of TOR. The deacetylase SIRT1 and the kinase AMPK are connected in a positive feedback loop concerning sensing low-energy states of cells.

Human aging is mostly associated with a progressive mitochondrial dysfunction coming with a decrease in NAD+ levels leading to impaired function of SIRT1 (Sect. 6.6). SIRT1 is one of the best-studied human signaling proteins regulating the metabolism of glucose and fat in response to energy level changes. Therefore, SIRT1 acts as a central control of the energy homeostasis network. Furthermore, SIRT1 controls the activity of the co-activator protein PPARGC1A, which in turn manages central metabolic pathways counteracting aging, such as mitochondrial biogenesis, enhanced anti-oxidant defenses and improved fatty acid β-oxidation. Thus, **anabolic signaling accelerates aging, while decreased nutrient signaling extends the lifespan**.

6.3   Principles of Insulin Signaling

Insulin resistance is a major predictor for the development of T2D (Chap. 9) and the metabolic syndrome (Chap. 10). In order to develop new drugs for the treatment of T2D and its cardiovascular complications, it is essential to understand the principles of insulin signaling. The major role of insulin signaling is the regulation of glucose, lipid and energy homeostasis, predominantly *via* actions on skeletal muscle, liver and WAT. **Final outcomes of insulin signaling are increased glucose uptake in muscle and fat tissue and inhibition of glucose synthesis in the liver**. In adipose tissue, insulin also inhibits the release of FFAs (free fatty acids).

Insulin acts *via* the IR being located in the plasma membrane of insulin sensing cells. When insulin binds to the α-subunit of the tetrameric receptor complex (two α- and β-subunits), it undergoes a conformational change that activates the cytosolic kinase domain of the β-subunit and allows the recruitment of IRS proteins to its cytosolic component. The activity of IR is upregulated by tyrosine phosphorylation, while the receptor is negatively regulated by protein tyrosine phosphatases. Moreover, SOCS (suppressor of cytokine signaling) 1 and 3, GRB10 (growth factor receptor-bound protein 10) and ENPP1 (ectonucleotide pyrophosphatase/ phospho-diesterase 1) inactivate the function of IR by blocking its interaction with IRS proteins or by modifying the receptor's kinase activity. SOCS proteins are upregulated in insulin resistance (Sect. 9.2).

In particular during insulin resistance IR is inactivated by ligand-stimulated internalization and degradation. IR has at least 11 intracellular substrates, such as IRSs 1-6, GAB1 (GRB2-associated binder 1), CBL (Cbl proto-oncogene, E3 ubiquitin protein ligase) and different isoforms of SHC (Src homology 2 domain containing) protein. The interaction of phosphorylated IRS proteins with the regulatory subunit of the kinase PI3K results in the generation of the second messenger PIP3 (phosphatidylinositol-3,4,5-triphosphate) activating the kinase AKT (Fig. 6.3). **IR/ IRS, PI3K and AKT form three important nodes in the insulin signal transduction pathway** being responsible for most of the metabolic actions of insulin, such as glucose uptake, glucose synthesis and inhibition of gluconeogenesis. In this way,

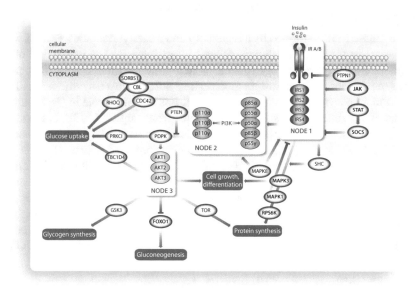

Fig. 6.3 Critical nodes in insulin signaling. Three important nodes of the insulin signaling network are IR/IRS, PI3K with its several regulatory and catalytic subunits and the three isoforms of AKT. Downstream of these nodes are the proteins TBC1D4 (TBC1 domain family member 4), PRK (protein kinase) C1, sorbin and SORBS1 (SH3 domain containing 1), CDC42 (cell-division cycle 42), GSK3 (glycogen synthase kinase 3), RPS6K (ribosomal protein S6 kinase), PDPK (3-phosphoinositide dependent protein kinase) 1 and 2, PTEN (phosphatase and tensin homologue), PTPN1 (protein tyrosine phosphatase, non-receptor type 1), RHOQ (ras homolog family member Q), CBL, JAK, FOXO1, TOR, SHC, SOCS, STAT, MAPK (mitogen-activated protein kinase) 1, 3 and 8

the three nodes explain the diversification and fine-tuning of insulin signaling both in health and disease.

IRS1 and IRS2 are ubiquitously expressed, while IRS3 is found primarily in adipocytes and the brain and IRS4 in embryonic tissues. After IR phosphorylates IRS' tyrosine sites the protein interacts with the SH2 (Src homology 2) domain-containing adaptor proteins, such as the regulatory subunits of PI3K or GRB2. GRB2 then associates with the adaptor protein SOS (son of sevenless) to activate the MAPK pathway *via* MAPK1, MAPK3 and MAPK8. This links the action of insulin to the control of cell growth and differentiation. Like for IR, the signaling function of IRS proteins is also regulated by the action of tyrosine phosphatases, such as SHP2 (SH2-domain-containing tyrosine phosphatase 2). SHP2 dephosphorylates the IRS binding sites of PI3K and GRB2 and interrupts their respective signal transduction pathways. In response to insulin, FFAs and cytokines IRS proteins are phosphorylated at serine residues. Most of these serine phosphorylations negatively regulate IRS signaling, *i.e.*, they represent a negative-feedback mechanism for the insulin signaling. Interestingly, serine phosphorylation of IRS1 strongly correlates with insulin resistance (Sect. 9.2). Moreover, reduced expression of IRS proteins or IR contributes to insulin resistance.

The kinase PI3K is formed by a regulatory and a catalytic subunit, each of which has several isoforms. The catalytic subunit is activated *via* the interaction of two SH2 domains in the regulatory subunit of IRS proteins. Inhibition of the enzyme blocks most of insulin's actions on glucose transport, glycogen synthesis, lipid synthesis and adipocyte differentiation, *i.e.*, **PI3K has a central role in the metabolic actions of insulin**. The most important downstream target of PI3K is the serine/threonine kinase AKT, which phosphorylates a number of key proteins, such as GSK3, TBC1D4 and FOXO1. AKT has three isoforms, of which AKT2 is most important in controlling metabolic functions. Phosphorylation of GSK3 decreases the kinases' inhibitory activity on GS leading to increased glycogen synthesis, but GSK3 has also a number of additional targets. Activated TBC1D4 stimulates small GTPases that are involved in cytoskeletal re-organization, which is required for the translocation of the glucose transporter GLUT4 to the plasma membrane, *i.e.*, TBC1D4 controls glucose uptake. Finally, **AKT controls the activity of FOXO1** (Sect. 6.4).

6.4   Central Role of FOXO Transcription Factors

The FOX transcription factor family contains over 100 members, some of which are crucial for the regulation of metabolism. The proteins FOXO1, FOXO3, FOXO4 and FOXO6 form a subclass of the family. FOXO1 is highly expressed in organs that control glucose homeostasis, such as in liver, skeletal muscle and adipose tissue, as well as in β cells. FOXOs are activated by oxidative stress and ER stress *via* MAPK8 and are negatively regulated by the insulin signaling pathway *via* PI3K and AKT (Fig. 6.4a). In *C. elegans* and other model organisms the IR-PI3K-AKT-FOXO signaling axis shows central functions in aging (Sect. 6.1). **Optimal FOXO signaling ensures longer lifespan, while the de-regulation of this pathway contributes to the age-related diseases cancer and T2D**. This provides FOXO with a central role at the interconnection of aging and disease and suggests that the main function of FOXOs is to maintain homeostasis in response to environmental stress, such as an increased oxidative stress, starvation or overnutrition. In the liver, FOXO1 is activated during starvation or low glucose levels. Under these conditions the transcription factor changes metabolism so that it ensures glucose homeostasis, *i.e.*, it initiates the breakdown of glycogen and gluconeogenesis. At moderate insulin resistance, reduced insulin signaling results *via* FOXO-stimulated gluconeogenesis in hyperglycemia. At severe insulin resistance, high levels of FOXO-triggered lipid oxidation lead to ketoacidosis. FOXO1 also affects the fasting response *via* actions in the CNS, such as in AGRP (agouti-related peptide) neurons (Sect. 8.4). Thus, **a balanced regulation of FOXOs *via* the IR-IRS-PI3K-AKT axis is essential for normal transcriptional control during the metabolic response**.

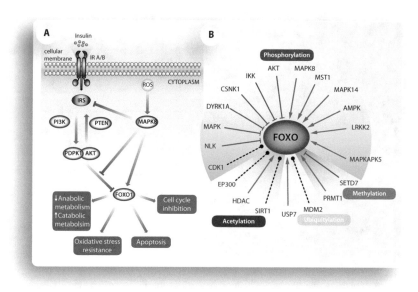

Fig. 6.4 Control and outcome of FOXO activation. Antagonistic control of FOXO through PI3K-AKT and MAPK8 signaling (**a**). Insulin and growth factors signal converge in their signaling at PI3K and activate PDPK1 and AKT. This is counteracted by PTEN. Active AKT can inhibit the transcriptional activity of FOXOs through phosphorylation at three conserved residues, resulting in cytoplasmic retention of FOXO. Oxidative stress induces the nuclear translocation of FOXOs *via* MAPK8 and thereby activates FOXO target genes. MAPK8 can inhibit the action of insulin at multiple levels and thereby counteracts the inhibition of FOXO. The activation of FOXO inhibits the cell cycle, increases oxidative stress resistance, induces apoptosis and shifts cellular metabolism away from anabolic pathways towards catabolic metabolism. FOXOs are regulated *via* post-translational modifications, such as phosphorylation, acetylation, methylation and ubiquitylation, by various signal transduction pathways (**b**)

FOXOs are regulated by a large number of signal transduction pathways and post-translational modifications (Fig. 6.4b). This is similar to other intensively studied transcription factors, such as p53, and suggests that combinatorial codes of post-translational modification regulate the function of key transcription factors. For example, the meaning of FOXO acetylation may be to switch FOXO-induced gene expression from an apoptotic to a pro-survival response. Such a code is analogous to the histone code (Box 5.1). FOXOs interact with SIRT1 during oxidative stress, but only when active PI3K signaling also promotes uptake of SIRT1 into the nucleus. Moreover, the energy sensor AMPK (Sect. 6.6) phosphorylates FOXOs, *i.e.*, AMPK directs the transcriptional action of FOXOs. In this way, AMPK activates alternative energy sources and stress resistance, as it is observed under conditions of calorie restriction (Sect. 6.5). Interestingly, AMPK activates FOXOs without influencing its subcellular localization, but it affects the shuttling of FOXO co-regulators. Thus,

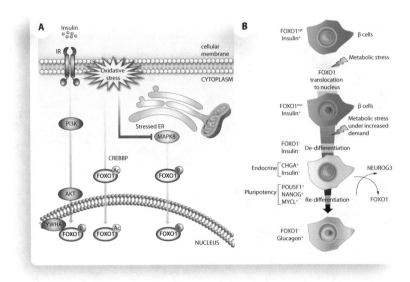

Fig. 6.5 FOXO1 signaling is affected by metabolic and oxidative stress. Continuous insulin stimulation activates AKT, which in turn phosphorylates serine and threonine sites of FOXO1, so that the transcription factor is retained to the cytoplasma (**a**). However, under oxidative stress conditions FOXO1 is acetylated by CREBBP and locates preferentially in the nucleus. The acetylation prevents the ubiquitination of FOXO1, *i.e.*, the protein is stabilized and retained in the nucleus. In addition, oxidative stress also activates MAPK8, which phosphorylates FOXO1 at serine and threonine sites distinct from those phosphorylated by AKT and further enhances nuclear retention of FOXO1. Furthermore, also ER stress can induce nuclear retention of FOXO1 *via* MAPK8 activation. Under severe hyperglycemic conditions, β cells no longer express FOXO1, insulin, PDX1 and MAFA but the pluripotency markers POU5F1, NANOG and MYCL, indicating de-differentiation into progenitor-like cells (**b**). Some of these cells start to express glucagon suggesting they re-differentiate into α cells

FOXO transcription factors function as scaffolds that are post-transcriptionally modified by a number of common signal transduction pathways.

Under conditions of tissue homeostasis FOXOs are inactive and located in the cytoplasma (Fig. 6.5A). The signal transduction of both insulin and growth factors converge at PI3K, the activation of which leads to an increased level of PIP3. This second messenger then activates PDPK1 (3-phosphoinositide dependent protein kinase 1), which in turn stimulates AKT. Active AKT translocates to the nucleus and phosphorylates FOXOs at three conserved residues, resulting in increased binding of the transcription factors to YWHA (tyrosine 3-monooxygenase/tryptophan 5-monooxygenase activation protein, also called 14-3-3). YWHA proteins bind more than 200 functionally diverse signaling proteins, such as transcription factors, kinases and transmembrane receptors, when they are phosphorylated at serine or threonine residues and retain them in inactive form in the cytoplasma. In contrast, under cellular stress, in particular at high ROS levels, FOXOs translocate into the nucleus and activate their target genes (Fig. 6.5a). MAPK8 counteracts the activity of insulin at multiple levels, such as by decreasing IRS activity (Sect. 6.3) and by inducing the release of FOXOs from YWHA proteins.

In mild hyperglycemia, *i.e.*, at metabolic stress, FOXO1 translocates into the nucleus of β cells. Under condition of additional stress, such as severe hyperglycemia, FOXO1 is downregulated and the cells loose the expression of insulin and the transcription factors PDX1 (pancreatic and duodenal homeobox 1) and MAFA (v-maf avian musculoaponeurotic fibrosarcoma oncogene homolog A) (Fig. 6.5b). Then β cells change their transcriptional profile and instead express the endocrine progenitor markers CHGA (chromogranin A) and NEUROG3 (neurogenin 3) and the pluripotency markers POU5F1 (POU class 5 homeobox 1), NANOG (Nanog homeobox) and MYCL (v-myc avian myelocytomatosis viral oncogene lung carcinoma derived homolog). **This is one mechanism of β cell failure** (Sect. 9.3)**, in which the cells de-differentiate into progenitor-like cells and re-differentiate into glucagon-producing α cells**.

6.5   Calorie Restriction from Yeast to Mammals

For thermodynamic reasons life does not function without energy supply, *i.e.*, there would be no life without high-energy nutrients, such as fatty acids or glucose (Sect. 1.2). When lowering food intake of model organisms, such as yeast, worms or flies, their lifespan rises to a maximum but then declines rapidly. This indicates that there is a species-specific optimal percentage of dietary reduction and an amplitude of response in terms of life extension (Fig. 6.6). In general, calorie restriction inactivates one or several nutrient signaling pathways, such as those of insulin/IGF1 or TOR. During periods of food scarcity the organisms enter a standby mode, in which cell division and reproduction are minimized or even stopped, in order to save energy for maintenance systems putting survival in preference before reproduction. Thus, **most species have developed an anti-aging system, in order to overcome periods of starvation**.

In humans and other mammals, extreme calorie restriction causes detrimental health effects, such as infertility and immune deficiencies. However, beneficial effects of moderate calorie restriction are obtained both by reducing carbohydrate intake as well as by reducing fat or protein consumption, *i.e.*, also in mammals several nutrition sensing pathways respond to reduction in food intake. Calorie restriction can increase the lifespan of rodents by up to 60%. In general, dietary-restricted rodents show many levels of metabolic, hormonal and structural adaptations when reducing body fat mass, such as higher insulin sensitivity, reduced inflammation and oxidative damage (Table 6.1). This is also observed in calorie-restricted monkeys. For example, in rhesus monkeys a 30% calorie restriction over 20 years reduced the incidence of cancer and CVDs by 50% compared with controls. In analogy, also in humans, calorie restriction provides beneficial effects against obesity, insulin resis-

		Life-span increase [fold]		Beneficial health effects	
		Dietary restriction	Mutations/ drugs	Dietary restriction	Mutations/ drugs
	S. cerevisiae	3	10 with starvation	Extended reproductive period	Extended reproductive period, decreased DNA damage mutations
	C. elegans	2-3	10	Resistance in misexpressed toxic proteins	Extended motility, resistance in mis-expressed toxic proteins and germ-line cancer
	D. melanogaster	2	1.6-1.7	None reported	Resistance to bacterial infection, extended ability to fly
	M. musculus	1.3-1.5	1.3-1.5 100% in combi-nantion with dietary restriction	Protection against cancer, diabetes, atherosclerosis, cardiomyopathy, autoimmune, kidney and respiratory diseases, reduced neurodegeneration	Reduced tumor incidence; protection against age-dependent cognitive decline, cardiomyopathy, fatty liver and renal lesions, extended insulin sensitivity
	M. mulatta	trend noted	not tested	Prevention of obesity, protection against diabetes, cancer and CVDs	not tested
	H. sapiens	not tested	not tested GHR-deficient subjects reach old age	Prevention of obesity, diabetes, hypertension reduced risk factors for cancer and CVDs	Possible reduction in cancer and diabetes

Fig. 6.6 Calorie restriction. In a number of model organisms experiments of calorie restriction have been performed, where nutrient-sensing pathways have been modulated genetically or chemically. However, there is a wide range of results and the long-term effects in human are not yet known

Table 6.1 Effects of calorie restriction on mammalian tissues

Tissue	Effects on dietary restriction
Liver	Increase in gluoconeogenesis and glycolysis
	Decrease in glycolysis
Muscle	Increase in mitochondrial biogenesis and respiration
	Increase in β-oxidation of fatty acids
	Increase in protein turnover
Fat	Decrease in storage of triacylglycerols
	Decrease in secreted leptin
	Increase in secreted adiponectin
Pancreatic β cells	Decrease in secreted insulin
Brain	Decrease in pituitary secretion of growth hormone, thyroid hormone, gonadotropins
	Increase in adrenal secretion of corticoids
Whole organism	Increase in insulin sensitivity and decrease in blood glucose
	Increase in metabolism

tance, inflammation and oxidative stress. Moreover, humans also show adaptations in their hormonal circuits, such as increased levels of adiponectin and reduced concentrations of the hormones triiodothyronine, testosterone and insulin. Moreover, reduced concentrations of cholesterol and CRP are observed and also blood pressure decreases. Thus, **in rodents, monkeys and even in humans calorie restriction protects against age-related diseases, such as T2D, CVDs and cancer**.

Despite convincing evidence of health benefits from calorie restriction, in the general population this approach may not be socially and ethically accepted for reducing the risk of age-related diseases, *i.e.*, for increasing the healthspan. An alternative may be repeated fasting and eating cycles, which, at least in rats, prolonged the lifespan by more than 80%. Moreover, mice that got a high-fat diet with regular fasting breaks were lean, had lower levels of circulating inflammatory markers and no fatty liver compared to mice that consumed an equivalent total number of calories *ad libitum*. There are different strategies in altering energy intake or the duration of fasting and feeding periods that can improve the health in mammals, such as

- classical caloric restriction, *i.e.*, a daily decrease of the recommended intake by 15–40%
- time-restricted feeding, *i.e.*, daily food intake within a 4–12 h window
- intermittent, periodic full or partial fasting, *i.e.*, a periodic, full- or multiday decrease in food intake
- fasting-mimicking diets; *i.e.*, reducing caloric intake and modifying diet composition but not fasting

Cycles of fasting or fasting-mimicking diets and refeeding promote the activation of HSCs (hematopoietic stem cells) and the regeneration of immune cells, modulate the gut microbiome and promote the T cell-dependent elimination of cancer cells.

Our body is genetically still largely adapted to a life as hunter and gatherer, where the feeding pattern was most likely shifting between starvation and times of overeating after a successful hunt. However, the efficacy of most fasting strategies are probably limited, if they are not combined with diets that have health-associated benefits, such as Mediterranean or Nordic diet (Box 1.1). Thus, **the switch between nutrient intake, usage and storage, *i.e.*, between feeding and fasting, is a fine-tuned regulatory, evolutionarily conserved program that involves the nutrient-sensing insulin/IGF1 and TOR signaling pathways and the food restriction pathways involving sirtuins and AMPK**.

6.6 Cellular Energy Status Sensing by Sirtuins and AMPK

The combination of nutritional epigenetics (Chap. 5) and beneficial health effects of calorie restriction and feeding-fasting alternations (Sect. 6.5) suggests that proteins, such as the HDACs of the sirtuin family, may be the key molecules in this nutrige-

Table 6.2 Mammalian sirtuins. The subcellular localization of sirtuins, their mode of action and their functions in different compartments are listed

Sirtuins	Molecular mass [kDa]	Cellular localization	Activity	Key regulatory functions
SIRT1	81.7	Nucleus and cytosol	Deacetylase	Metabolism, inflammation
SIRT2	43.2	Cytosol	Deacetylase	Cell cycle and motility, myelination
SIRT3	43.6	Mitochondria	Deacetylase	Fatty acid oxidation, antioxidant defences
SIRT4	35.2	Mitochondria	ADP-ribosyl-transferase	Amino acid-stimulated insulin secretion, suppression of fatty acid oxidation
SIRT5	33.9	Mitochondria	Deacetylase, Demalonylase, Desuccinylase	Urea cycle
SIRT6	39.1	Nucleus	Deacetylase, ADP-ribosyl-transferase	Genome stability, metabolism
SIRT7	44.8	Nucleus	Deacetylase?	Ribosomal DNA transcription

nomic process. **Sirtuins combine epigenetically important enzymatic activity with the sensing of the energy status of cells**, *i.e.*, the NAD^+/NADH ratio. In humans the sirtuin family has seven members, SIRT1 to SIRT7, that act in different cellular compartments (Table 6.2). In the nucleus, the substrates of sirtuins are histones and transcription factors, but also in the cytoplasm and in mitochondria they remove acetyl groups from post-translationally modified regulatory proteins. SIRT1, SIRT2, SIRT3, SIRT6 and SIRT7 are specific protein deacetylases, whereas SIRT4 and SIRT5 preferentially remove other acyl groups from lysines. SIRT1, SIRT6 and SIRT7 are predominantly localized in the nucleus but at least SIRT1 is also found in the cytosol. In contrast, SIRT2 is mainly cytosolic and enters the nucleus only during the G2 to M phase transition of the cell cycle. SIRT3, SIRT4 and SIRT5 are preferentially located in mitochondria.

In a simplified view, the deacetylase activity of sirtuins counterbalances nutrient-driven protein acetylation. **During fasting and/or exercise, levels of the sirtuin co-factor NAD^+ rise in skeletal muscle, liver and WAT, while high-fat diet reduces the NAD^+/NADH ratio**. Pharmacological targeting of sirtuins, in particular of SIRT1, is promising for the treatment of T2D. The search for further natural or synthetic sirtuin activators led to the identification of several compounds, of which resveratrol, a polyphenol found in red grapes and berries (Fig. 5.1), received most attention. Resveratrol improves mitochondrial activity and metabolic control in humans, *i.e.*, it may also increase our healthspan. The most prominent synthetic sirtuin activator is SRT2104, which protects against diet-induced obesity by improving mitochondrial function. However, both resveratrol and SRT1720 do not activate sirtuins directly, but at least resveratrol acts *via* AMPK.

AMPK is activated by stress that inhibits the catabolic production of ATP, such as starvation for glucose or oxygen, as well as by stress that increases ATP consumption, such as muscle contraction (Box 6.2). Furthermore, numerous synthetic compounds, such the anti-diabetic drugs metformin, phenformin and thiazolidinediones, as well as plant products, such as resveratrol, epigallocatechin gallate (green tea), capsaicin (peppers) and curcumin (turmeric) activate AMPK. However, the activation by most of these compounds, including metformin and resveratrol, is indirect *via* the increase of cellular AMP and ADP through the inhibition of mitochondrial ATP synthase. Since these AMPK activators also extend the lifespan of *C. elegans*, they may act as mimetics of calorie restriction (Sect. 6.5) and/or exercise (Sect. 1.6). Both processes decrease the cellular energy status and have beneficial effects on the healthspan.

Box 6.2: Regulation of AMPK

ATP is mainly generated by oxidative phosphorylation in the mitochondrial membrane. A high ATP/ADP ratio is essential to promote nearly all processes in living cells, since they all require energy and mostly are driven by the hydrolysis of ATP to ADP. Then, approximately every second ADP molecule is converted by the enzyme adenylate kinase to AMP. Therefore, falling cellular energy is associated with a decreased ATP/ADP ratio on one side and with increases in both ADP and AMP on the other side. Thus, **the energy status of a cell can be monitored either *via* the ATP/ADP or the ATP/AMP ratio**. The ATP/AMP ratio is directly sensed by a few enzymes, such as glycogen phosphorylase and phosphofructokinase (both get activated) and fructose-1,6-bisphosphatase (which gets inhibited). However, **the main sensor of the energy status of cells is AMPK**. AMPK is a heterotrimeric complex of its catalytic α-subunit and the regulatory β- and γ-subunits and is activated by various types of metabolic stresses, drugs and xenobiotics involving increases in cellular AMP, ADP or Ca^{2+}. The major upstream kinases of AMPK are LKB1 (liver kinase B1) and CAMKK2 (Ca^{2+}/calmodulin-dependent protein kinase kinase 2). LKB1 provides a high basal level of AMPK phosphorylation that is modulated by the AMP binding, while the alternative activation pathway *via* CAMKK2 responds to increases in cellular Ca^{2+} but is independent of AMP or ADP level changes.

AMPK activation has multiple effects on cellular metabolism (Fig. 6.7). In general, the activation of AMPK by lack of energy stimulates catabolic pathways that generate ATP, while it turns off anabolic pathways that consume ATP. The AMPK-controlled catabolic processes include

- upregulation of glucose uptake by promoting the expression and function of glucose transporters

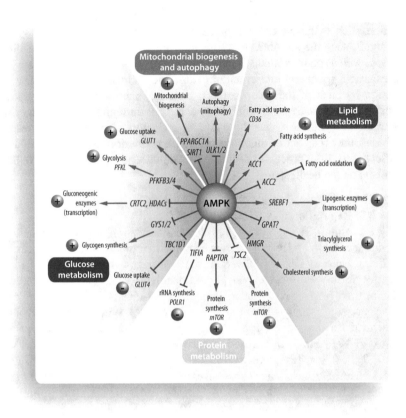

Fig. 6.7 Consequences of AMPK activation. This scheme displays the metabolic effects of AMPK activation. Proteins shown with question marks may not be directly phosphorylated by AMPK. *CRTC2* CREB-regulated transcription co-activator 2, *GPAT* glycerol phosphate acyl transferase, *RAPTOR* regulatory associated protein of TOR, *TBC1D* TBC1 domain, *TIFIA* transcription initiation factor IA; *TSC2* tuberous sclerosis 2

- promotion of glycolysis under anaerobic conditions by phosphorylating and activating PFKFB2 (6-phosphofructo-2-kinase/fructose-2,6-biphosphatase 2), the enzyme responsible for the synthesis of glycolytic activator fructose 2,6-bisphosphate
- stimulating mitochondrial biogenesis
- inhibiting gluconeogenesis
- reducing glycogen synthesis *via* the inhibition of GYS (glycogen synthase).

Furthermore, AMPK increases fatty acid synthesis by stimulating ACC (acetyl-CoA carboxylase) 1, the key regulatory enzyme in fatty acid synthesis. Moreover, AMPK upregulates the expression of enzymes involved in fatty acid synthesis and inhibits the lipogenic transcription factor SREBF1. Furthermore, AMPK promotes uptake and fatty acid β-oxidation in mitochondria *via* inhibition of ACC2. Mitochondrial biogenesis is another important process activated by AMPK and is mediated *via* SIRT1 and PPARGC1A as well as *via* increased clearance of dysfunctional mitochondria *via* ULK1 (UNC-51-like kinase 1). Finally, this generates increased capacity for the oxidative catabolism of both fatty acids and glucose. AMPK also conserves ATP by inactivating anabolic pathways, such as the biosynthesis of lipids, carbohydrates, proteins and ribosomal RNA. AMPK can downregulate the expression of the proteins involved in these pathways. Moreover, AMPK also influences whole-body metabolism and energy balance *via* mediating effects of hormones and other agents acting on neurons of the primary appetite control center, the arcuate nucleus of the hypothalamus, that regulates intake of food and energy expenditure actions (Sect. 8.4).

The circadian clock is a conserved mechanism that allows anticipating and responding to environmental changes, such as feeding and fasting (Sect. 3.6). Since there is a strong relation between the circadian clock and metabolism, **time-restricted feeding can restore the cycling of metabolic regulators**. Epigenetics of the circadian clock coordinates daily behavioral cycles of sleep-wake and fasting-feeding with anabolic and catabolic processes in the periphery. Thus, NAD^+ oscillation, redox flux, ATP availability and mitochondrial function are used in order to influence acetylation and methylation reactions of chromatin modifiers (Sect. 5.2). Moreover, gene expression changes related to dietary restriction promote global preservation of genome integrity and chromatin structure, such as maintenance of heterochromatin (Fig. 6.8). Since nutrition signaling pathways *via* insulin, TOR and NAD^+-sensing sirtuins integrate metabolic signals into chromatin responses, they inform the epigenome on nutrient availability and thus have a key role in determining lifespan. In contrast, artificial light, shift work, travel and temporal disorganization disrupt the alignment between the external light-dark cycle and their internal clock and cause a disadvantage for metabolic health (Sect. 3.6).

In humans physical fitness and longevity are strongly associated, since regular exercise reduces morbidity and mortality (Sect. 1.6). Moreover, physical activity promotes healthy aging *via* preventing cognitive decline. It induces changes in the chromatin of skeletal muscles, such as increased H3K36ac levels and the cellular localization of HDAC4 and HDAC5. Thus, **physical activity has direct effects on the epigenome promoting healthy aging**.

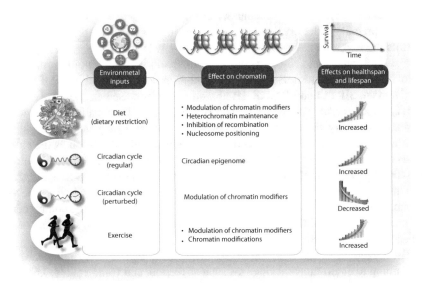

Fig. 6.8 Effects of environmental inputs on chromatin and longevity. Many environmental signals that modulate lifespan also affect chromatin. These are dietary restriction, the circadian cycle and physical activity. More details are provided in the text

Additional Readings

Campisi J, Kapahi P, Lithgow GJ, Melov S, Newman JC, Verdin E (2019) From discoveries in ageing research to therapeutics for healthy ageing. Nature 571:183–192

Di Francesco A, Di Germanio C, Bernier M, de Cabo R (2018) A time to fast. Science 362:770–775

Haeusler RA, McGraw TE, Accili D (2018) Biochemical and cellular properties of insulin receptor signalling. Nat Rev Mol Cell Biol 19:31–44

Lopez-Otin C, Galluzzi L, Freije JMP, Madeo F, Kroemer G (2016) Metabolic control of longevity. Cell 166:802–821

Singh PP, Demmitt BA, Nath RD, Brunet A (2019) The genetics of aging: a vertebrate perspective. Cell 177:200–220

Steinberg GR, Carling D (2019) AMP-activated protein kinase: the current landscape for drug development. Nat Rev Drug Discov 18:527–551

Chapter 7
Chronic Inflammation and Metabolic Stress

Abstract In this chapter, we will present monocytes and macrophages as the key players in acute and chronic inflammation. Macrophages react stimulus- and tissue-specifically and either develop from monocytes that are circulating in the blood or from self-renewing embryonal cell populations. M1-type macrophages are key cells in the initiation of the acute inflammatory response, while M2-type macrophages are resolving inflammation and coordinate tissue repair. Tissue inflammation is not only caused by bacterial infection or tissue injury but also derives from changes in the concentration of nutrients and metabolites. We will provide examples of metabolic stress, such as disturbance of reverse cholesterol transport and ER stress. The latter is, in contrast to infectious or traumatic stress, often caused by lipid overload in the blood and in adipose tissue. Thus, **the immune system is implicated in the regulation of metabolic homeostasis, while perturbations in this immune-metabolic network are often the basis of the different features of the metabolic syndrome**.

Keywords Monocytes · M1- and M2-type macrophages · Dendritic cells · Cytokines · Acute inflammation · Chronic inflammation · Inflammasome · Reverse cholesterol transport · Metabolic stress · ER stress · Unfolded protein response

7.1 The Central Role of Monocytes and Macrophages

Monocytes constitute 1–5% of circulating leukocytes and are produced in the bone marrow from common myeloid progenitors of granulocytes (*i.e.*, primarily neutrophils) and monocytes, referred to as M-CFUs (colony-forming units) (Fig. 7.1). The differentiation of myeloid progenitors is orchestrated by pioneer transcription factors, such as SPI1 and CEBPA, in cooperation with changes of the epigenome (Chap. 5). After their differentiation, monocytes are released into the blood stream and migrate through the endothelium of blood vessels into those tissues that send out activating signals. This movement is mediated by the cytokine- and chemokine-induced expression of adhesion molecules, such as selectins, on the surface of endo-

© Springer Nature Switzerland AG 2020
C. Carlberg et al., *Nutrigenomics: How Science Works*,
https://doi.org/10.1007/978-3-030-36948-4_7

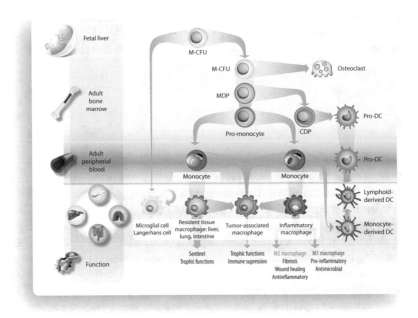

Fig. 7.1 Differentiation of monocytes. M-CFUs in the bone marrow are the precursors to MDPs (macrophages and DC progenitors). In the bone marrow, MDPs differentiate to CDPs (common DC progenitors) or to pro-monocyte precursors. Langerhans cells in the skin, microglial cells in the brain and a number of other tissue-resident macrophages initially develop during embryogenesis from M-CFUs in the yolk sac or fetal liver. The remaining tissue macrophages get polarized, depending on the inflammatory milieu, into M1- or M2-type macrophages (Fig. 7.2). Monocytes are also the precursors to DCs

thelial cells. Within tissues, monocytes differentiate into macrophages or DCs, *i.e.*, into the central cellular components of the innate immune system (Box 2.2). Thus, **a central role of monocytes is to refill the pool of macrophages and DCs in response to inflammation and other stimuli**.

Nutrient deprivation and infection by pathogens are the most challenging events for the survival of an organism. Thus, **there was a co-evolution for the responses to food *via* the endocrine regulation of metabolism and to infectious diseases *via* the innate immune system**. The interface between metabolism and immunity is largely mediated by macrophages. Tissue-resident macrophages, such as microglia and Langerhans cells, have a wide variety of homeostatic and immune surveillance functions ranging from the clearance of cellular debris, the response to infections and the resolution of inflammation. A substantial proportion of these macrophages is self-renewing and derives during embryogenesis from the yolk sac and the fetal liver (Fig. 7.1). Populations of tissue-resident macrophages are found

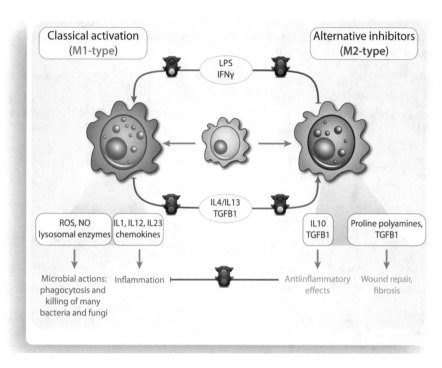

Fig. 7.2 Classical and alternative macrophage activation. Different stimuli activate mono-cytes/macrophages to develop into functionally distinct populations. Classically activated macro-phages (M1-type) are induced by cytokines, such as IFNγ, and microbial products, such as LPS. They are microbicidal and involved in potentially harmful inflammation. Alternatively acti-vated macrophages (M2-type) are induced by the cytokines IL4, IL13 and TGFB1, which are produced by T_H2 cells, and are important in tissue/wound repair and fibrosis

in most tissues of our body, such as Kupffer cells in the liver, microglia in the CNS, osteoclasts in the bone and alveolar macrophages in the lung. In response to their activation by pathogens or metabolites, macrophages secrete a number of signaling molecules, such as cytokines, chemokines and growth factors, which affect the migration and activity of other immune cells. This response is called acute inflam-mation, when it is caused by infection or injury, and is often associated with ery-thema, hyperthermia, swelling and pain. However, it resolves within a few days to weeks. In contrast, low-grade chronic inflammation, such as in obesity (Sect. 8.3), does not cause heat or pain, but it can last over months and years, when the origin of the stimulus cannot be resolved (Sect. 7.2).

Macrophages are often grouped into two classes, M1- and M2-type macro-phages, which represent two extremes of a continuum of functional profiles (Fig. 7.2). Pro-inflammatory molecules, such as INFγ (interferon gamma), TNF and TLR activators, such as LPS (lipopolysaccharide), induce M1-type macrophages that in turn secrete further pro-inflammatory molecules, in order to sustain the inflammatory reaction. This classical pathway of IFNγ-dependent macrophages activation provokes the adaptive immune system to respond through the proliferation of type 1 T_H cells (T_H1) (Box 7.1).

Box 7.1: Subsets of T Lymphocytes

T cells constitute up to 30% of all circulating leukocytes and are a major component of the adaptive immune system (Box 2.2). They occur in a number of important subtypes, such as T_H and cytotoxic T cells. T_H cells are characterized by the expression of the CD4 glycoprotein on their surface and support other cells of the immune system, such as maturation of B cells into antibody producing plasma cells and memory B cells as well as they activate of cytotoxic T cells and macrophages. Antigen-presenting cells, such as DCs, activate T_H cells *via* the presentation of microbe-derived peptides on MHC II receptors. Depending on their cytokine expression profile T_H cells are subdivided into T_H1 (IFNγ and IL2), T_H2 (IL4, IL5 and IL13) and T_H17 (IL17) cells. Moreover, T_H cells can differentiate into T_{REG} cells that are involved in immune tolerance and inhibit overboarding immune responses. Cytotoxic T cells (also called T killer cells) are characterized by CD8 expression and can destroy virus-infected cells and tumor cells. They get activated *via* the presentation of peptides (derived from intracellular proteins) on MHC I receptors on the surface of all nucleated cells.

In contrast, M2-type macrophages exert an almost opposite immuno-phenotype. They do not produce nitric oxide (NO) or radicals required for killing of microbes but provoke immunotolerance and T_H2-type immune responses. M2-type macrophages produce anti-inflammatory molecules, such as TGFB1 (transforming growth factor beta 1), IL10 or IL1RN (IL1 receptor antagonist), and inhibit the secretion of pro-inflammatory cytokines. This alternate macrophage pathway is induced by anti-inflammatory molecules, such as CSF2 (colony stimulating factor 2), TGFB1 and the T_H2-type cytokines IL4 and IL13, and nuclear receptor ligands, such as glucocorticoids. **The main role of M2-type macrophages is resolution, such as tissue repair, wound healing, angiogenesis and extracellular matrix deposition**. For example, M2-type macrophages can be induced by the nuclear receptor PPARγ and maintain adipocyte function, insulin sensitivity and glucose tolerance, which prevents the development of diet-induced obesity and T2D (Chaps. 8 and 9). However, when obesity-associated danger signals are sensed by the NLRP (NOD-like receptor protein) 3 inflammasome (Box 7.2), this protein complex serves as a molecular switch that let the adipose tissue-associated macrophages turn from an M2- to an M1-type phenotype.

Box 7.2: The NLRP3 Inflammasome

A large protein complex composed of NLRPs, the adaptor protein ASC (apoptosis-associated speck) and the pro-inflammatory CASP (caspase) 1 that is formed in response to the rise of the levels of a number of PAMPs (pathogen-associated molecular patterns) and DAMPs (damage-associated molecular patterns) (Fig. 7.3). Inflammasome activation requires a priming signal from PPRs, such as TLR4, and a second signal involving potassium ion efflux, lysosomal damage or ROS generation. Cholesterol crystals in lysosomes can provide this second signal, either as a result of phagocytosis of extracellular cholesterol crystals or *via* uptake of modified LDLs and free cholesterol released from LDLs. Inflammasome activation leads to the secretion of the pro-inflammatory cytokines IL1B and IL18.

7.2   Acute and Chronic Inflammation

Acute inflammatory responses to insults, such as injury and infection, are critical for organism's health and recovery. Infection or tissue damage is initially sensed by PRRs, such as TLRs, NLRs, RLRs (retinoic acid-inducible gene 1 (RIG1)-like helicase receptors), lectins and scavenger receptors, that bind to PAMPs and DAMPs (Fig. 7.4a). PRRs are evolutionarily highly conserved and their initial function was providing anti-microbial immunity and regulating autophagy (Box 3.1). In response to nutrient deprivation this group of receptors had been optimized, in order to increase inflammation and insulin resistance, both of which are a strategies to combat pathogens. Thus, **a variety of different molecular motifs and stimuli converge on a small number of sensing receptors that trigger the innate immune response, cause inflammation and induce an appropriate adaptive metabolic response**. PAMP- or DAMP-stimulated PRRs start signal transduction pathways that end in most cases in the activation of inflammation-associated transcription factors, such as NFκB and AP1. Major targets of these transcription factors are genes encoding for pro-inflammatory cytokines, such as *TNF* and *IL1B*, anti-microbial factors, such as *NOS2* (inducible nitric oxide synthase 2), and cell-recruiting chemokines, such as *CCL2* and *CCL5*. Another important component of the DAMP sensing system is the inflammasome (Box 7.2) that primarily enhances IL1B production and secretion (Fig. 7.3).

On the cellular level, the microbial challenge of resident macrophages stimulates the influx of cells from the blood, such as neutrophils and monocytes (as a source of inflammatory macrophages) (Sect. 7.1). Most inflammatory lesions are initially dominated by these monocyte-derived macrophages. The changed expression profile of the macrophages activates further cells of the innate immune system as well as the adaptive immune system. The early inflammatory response comprises a number of redundant components and is further amplified in cytokine-mediated feed-

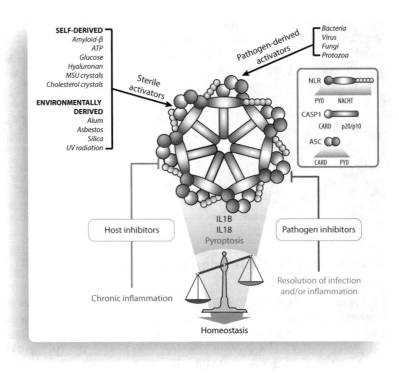

Fig. 7.3   Central role of the inflammasome. During acute or chronic inflammation, inflammasomes are directly or indirectly activated by a wide array of DAMPs and PAMPs. The initial event leads to the activation of CASP1, the release of IL1B and IL18 as well as sometimes to pyroptosis (a form of apoptosis associated with anti-microbial responses during inflammation). The release of IL1B and IL18 induces the recruitment of effector cell populations of the immune response and tissue repair, *i.e.*, the activation of inflammasomes results in the resolution of infection or inflammation and contributes to homeostatic processes. However, constant activation of the inflammasome can lead to chronic inflammatory diseases. Pathogen-derived inhibitors can block inflammasome activation and thus the resolution of infection, while host-derived inflammasome inhibitors will prevent chronic inflammation

forward loops. For example, adipose tissue of lean persons contains deactivated macrophages, eosinophils and T_{REG} cells, while within the initial phase of the acute inflammation there is rapid invasion of neutrophils that is followed by the recruitment of monocyte-derived macrophages, T cells and stromal cells. Thus, **under resting conditions most tissues contain only a few resident macrophages, while at acute inflammation the number of immune cells drastically increases and the characteristics of these cells changes.**

The combined action of innate and adaptive immune cells eradicates infectious microbes but also results in collateral tissue damage, such as cytotoxicity due to ROS production and the degradation of extracellular matrix *via* proteases. After the clearance of pathogens and the removal of apoptotic neutrophils by phagocytes, there is recruitment or phenotypic switching of macrophages into the M2-type, *i.e.*, into a pro-resolving phenotype (Fig. 7.2). This results in an overall repair and normalization of the tissue architecture and function, including the re-establishment of the vascularization. Most of these processes are activated by PUFAs and their derivatives, such as eicanosoids. These metabolites function as signaling molecules binding to membrane proteins, such as GPR120, or directly activate nuclear receptors and initiate intracellular pathways leading to an anti-inflammatory response (Chap. 3). This also involves anti-inflammatory cytokines, such as IL10 and TGFB1, and small lipid mediators, such as lipoxins, resolvins, protectins and maresins that are produced by the enzymes arachidonate 5- and 15-lipoxygenase (ALOX5 and ALOX15) from arachidonic acid and ω-3 fatty acids. Thus, **nutrition-derived signaling molecules are critical for resolving inflammation**.

In humans the basal inflammatory response increases with age, which is often referred to as "inflammaging", and leads to low-grade chronic inflammation that is maladaptive and further promotes the aging process. This may be due to

- the accumulation of senescent cells that secrete pro-inflammatory cytokines
- an increased likelihood that a failure of the immune system does not effectively clear pathogens and dysfunctional host cells
- overactivity of the transcription factor NFκB
- a defective autophagy response ultimately leading to increased ROS production.

In all these cases, not microbes but the excess of endogenous molecules, such as lipoproteins, SFAs or protein aggregates initiate the inflammatory response (Fig. 7.4b). Metabolic dysregulation associated with chronic inflammation accompanies not only aging itself but also most common age-related diseases, such as T2D and CVDs. Thus, **sterile (*i.e.*, non-bacterial) induction of low-grade chronic inflammation is a critical characteristic of aging as well as of metabolic diseases**. In both cases an enhanced activation of the inflammasome (Box 7.2) and other pro-inflammatory pathways leads to the increased production of IL1B, TNF and interferons. This inflammatory response amplifies the production of disease-specific DAMPs resulting in positive-feedback loops that accelerate the underlying disease process (Sect. 7.3). For example, inflammation stimulates the formation of oxidized phospholipids that can serve as DAMPs in atherosclerosis (Sect. 10.2). Thus, **an inhibition of chronic inflammation can reduce the rate of disease progression to a point of substantial clinical benefit**, although it does not alter the underlying pathogenic process and its cause.

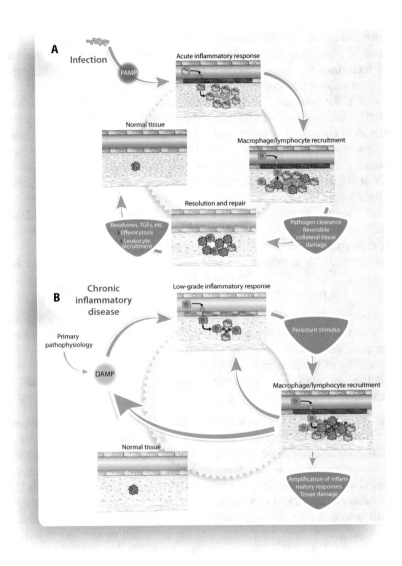

Fig. 7.4 Acute and chronic inflammation. Acute inflammatory response to infection is initiated by the presentation of PAMPs to PRRs (**a**). Eradication of the pathogen eliminates the stimulus but may cause some collateral but reversible tissue damage. The resolution/repair phase leads then to the restoration of normal tissue homeostasis. Chronic inflammation is caused by non-immune pathophysiological processes that trigger an initial sterile inflammatory response involving DAMPs and is amplified by cytokines and chemokines (**b**). However, this response does not eliminate the initial stimulus, so that non-resolving inflammation persists and results in continuous tissue damage

7.3 Reverse Cholesterol Transport and Inflammation

The protective role of macrophages against antigenic substances, such as pathogenic microbes, enables them to recognize also lipid molecules, such as cholesterol crystals, when these accumulate within tissues. This process links macrophages to metabolism and assigns the cells with critical roles in atherosclerosis and lipid storage diseases (Chap. 10). There are a number of interactions between inflammation and metabolism, such as that pathogens and inflammation alter metabolic processes as well as that metabolic diseases lead to abnormal immune reactions. A central example is the process of reverse cholesterol transport. The transporter proteins ABCA1 and ABCG1 in the membrane of cholesterol-laden macrophages interact with the main HDL protein APOA1 and mediate the efflux of cholesterol from the arterial wall to HDLs in the circulation (Fig. 7.5). The enzyme LCAT (lecithin-cholesterol acyltransferase) esterifies free cholesterol in HDLs forming cholesteryl esters. In the liver some of the free cholesterol or cholesteryl esters are selectively taken up *via* the transporter protein SCARB1 (scavenger receptor class B member 1) without degrading the HDLs. The liver either recycles the cholesterol as a component of secreted lipoproteins or excretes them as bile acids *via* the action of the transporters ABCG5 and ABCG8. In the plasma, the transporter protein CETP transfers cholesteryl esters from HDLs to VLDLs and LDLs in exchange for triacylglycerols. The enzymes LPL and LIPC (hepatic lipase) hydrolyze the triacylglycerols of VLDLs, so that primarily cholesteryl esters remain and the lipoproteins transform into LDLs.

Increased LDL-cholesterol levels in the circulation are the principal drivers of atherosclerosis (Sect. 10.2), since this causes cholesterol accumulation and inflammatory response in the artery wall (Fig. 7.5). Under healthy conditions, HDLs can oppose this process and are able to reduce inflammation by promoting the cholesterol efflux from foam cells. However, acute inflammation impairs reverse cholesterol transport at two central steps. This downregulates the expression of the genes *ABCA1* and *ABCG1* in macrophages, *i.e.*, it reduces cholesterol efflux from macrophages. In addition, it decreases the hepatic expression of the genes *APOA1, CETP, ABCG5, ABCG8* and *CYP7A1*, which results in the reduced excretion of cholesterol. Moreover, acute inflammation causes the accumulation of triacylglycerols within VLDLs leading to increased hepatic production of these lipoproteins and their reduced clearance from circulation by LPL. The increased VLDL levels maintain high lipid levels in peripheral tissues, in order to suppress infection and to allow tissue repair. Increased lipid levels may even be beneficial, at least for the short period of an acute microbe infection. However, during acute sepsis HDLs have a decreased ability to mediate cholesterol efflux from macrophages (Fig. 7.5). Under these conditions, HDLs do not act anymore as anti-inflammatory lipoproteins that suppress monocyte adhesion to endothelial cells but in contrast support the monocyte recruitment. Moreover, oxidized and aggregated LDLs can activate PPRs, such as TLRs, on the surface of macrophages and thus directly trigger the inflammatory

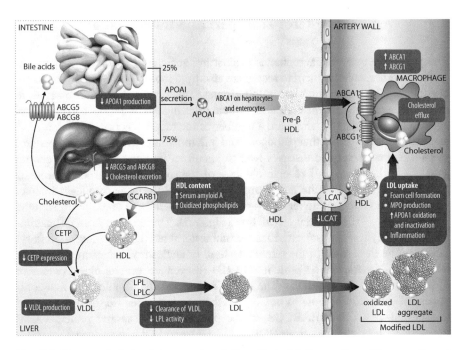

Fig. 7.5 Reverse cholesterol transport and the effect of inflammation. After secretion of APOA1 from the liver and the intestine, it interacts with ABCA1 on hepatocytes and enterocytes and is assembled into preβ-HDLs. In macrophages, these lipoproteins are loaded *via* the action of ABCA1 and ABCG1 with cholesterol and phospholipids and form HDLs. LCAT catalyzes the esterification of free cholesterol in HDLs. In the liver, SCARB1 removes some of the free cholesterol or cholesteryl esters from HDLs. In the liver, cholesterol is then either recycled into VLDLs *via* the action of CETP or is transformed into bile acids that are excreted by the transporters ABCG5 and ABCG8. In the blood stream, VLDLs undergo a lipolytic cascade mediated by the enzymes LPL and hepatic lipase and are transformed into cholesterol-rich LDLs. A small proportion of the LDLs end up in the arterial wall, where they are modified by oxidation or aggregation and are taken up by macrophages. This leads to the formation of macrophage foam cells, the production of MPO (myeloperoxidase) and inflammation. Red boxes indicated how inflammation negatively affects this reverse cholesterol transport

response. These modified LDLs are taken up and cause cholesterol accumulation within macrophages, which in turn amplifies the signal transduction downstream of TLRs. *Via* the increased production of cytokines and chemokines this boosts inflammation. Thus, **in a feed-forward mechanism the acute inflammation changes cellular cholesterol homeostasis, which further amplifies the inflammatory response**.

Chronic infections, such as HIV-1, and autoimmune diseases, such as systemic lupus erythematosus, rheumatoid arthritis and psoriasis, are often associated with reduced HDL-cholesterol levels, increased concentrations of atherogenic lipoproteins and accelerated atherosclerosis. Similar mechanisms may also contribute to other metabolic disorders. For example, on adipose tissue macrophages of obese persons TLRs and NLRs are activated by lipids, such as ceramides or SFAs. This can lead to chronic inflammation, insulin resistance and NAFLD (Sect. 9.2). During chronic inflammation, stimuli in atherosclerotic lesions, such as intracellular cholesterol crystals that activate the inflammasome (Fig. 7.3), induce the enzyme MPO that oxidizes APOA1 in HDLs. In this way, the lesions further compromise the capability of the lipoproteins concerning reverse cholesterol transport. In individuals with CHD, the levels of oxidized APOA1 inversely relate to the ABCA1 efflux capacity and positively correlate with atherosclerotic disease (Sect. 10.2). Therefore, the inflammatory response of macrophages leads to local inactivation of APOA1 in the arterial wall and reduces cholesterol efflux from macrophages. In turn, this causes cholesterol accumulation in macrophages and further enhances the inflammatory responses. Moreover, HDL-cholesterol levels decrease and their composition changes, and through the oxidation of APOA1 they become dysfunctional.

The inflammatory response in macrophages is primarily mediated *via* TLR3 and TLR4 and downregulates through the induction of the transcription factor IRF3 the expression of the LXR target genes *ABCA1* and *ABCG1* (Sect. 3.4). Although this reduces cholesterol efflux, promotes cholesterol accumulation and enhances the inflammatory response, the increased levels of cholesterol result in higher oxysterol concentrations and respective LXR activation. In this way, increased cholesterol levels can stimulate their own efflux. Thus, **the innate immune system can use changes in cholesterol metabolism, in order to amplify the inflammatory response, but also for restoring homeostasis**.

7.4 Sensing Metabolic Stress *via* the ER

Humans have a multitude of mechanisms of cellular adaptation to stress. Perturbations that induce cell stress include

- infectious agents, which can drive a large set of stress responses by activating PRRs (Sect. 7.2)
- nutrient deprivation, which activates autophagy in most cells, hence enabling them to catabolize their own components for survival (Sect. 3.1, Box 3.1)
- hypoxia, respiratory poisons and xenobiotics that cause mitochondrial stress
- DNA-damaging agents, such as ionizing radiation and some xenobiotics, that activate repair pathways
- heat shock or chemical toxins that cause protein denaturation, both of which activate the unfolded protein response in the ER.

The ER is an organelle that has a vital role in maintaining cellular and whole body metabolic homeostasis. It is the primary site of the 3D folding of all membrane proteins and secreted proteins that are produced in the cell and also performs their quality control. Although the ER has significant adaptive capacity to manage the periodic cycles associated with feeding, fasting and other metabolic demands of

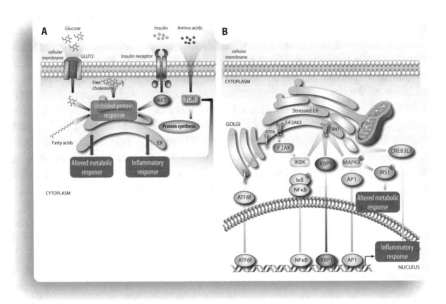

Fig. 7.6 The unfolded protein response. The ER responds to multiple nutrient-associated signals, such as those induced by fatty acids, glucose, free cholesterol, insulin and amino acids (**a**).
ER stress due to nutrient overload induces the unfolded protein response, which activates inflammatory signaling pathways that result in altered metabolic and inflammatory responses. The three
ER transmembrane proteins EIF2AK3, ERN1 and ATF6 are the molecular mediators of the
unfolded protein response (**b**). ERN1 activates *via* the phosphorylation of the kinases IKBK and
MAPK8 the transcription factors NFκB and AP1, which lead to an increase in the expression of
inflammatory genes. This response is further enhanced by translocation of ATF6 to the Golgi and
processing there to an active transcription factor, upregulation of the expression of the transcription
factor XBP1 and the transcription factor CREB3L3. MAPK8 also phosphorylates IRS1 resulting
in an altered metabolic response. A functional and molecular integration between the different
organelles can mediate the spread of the stress

limited duration, it is less flexible to respond appropriately to chronic and escalating
metabolic challenges. Thus, **an accumulation of unfolded proteins in the ER
lumen triggers an adaptive, protective mechanism, referred to as unfolded protein response** (Fig. 7.6a). For example, hepatocytes and adipocytes of obese human
show increased ER stress compared with lean controls. Moreover, multiple CVD
risk factors, such as inflammation, dyslipidemia, hyperhomocysteinemia and insulin resistance, can lead to the development of ER stress in atherosclerotic lesions.

On the molecular level, the ER stress response is primarily mediated by the proteins EIF2AK3 (eukaryotic translation initiation factor 2-alpha kinase 3), ERN1
(endoplasmic reticulum to nucleus signaling 1) and ATF6 (activating transcription
factor 6). In the absence of a stress signal, these three transmembrane proteins are
bound by the chaperone HSPA5 and are kept inactive. An increased protein load, in

particular improperly folded proteins, activates EIF2AK3 and ERN1, *i.e.*, they dissociate from HSPA5 and initiate signal transduction pathways. EIF2AK3 phosphorylates EIF2A (eukaryotic translation initiation factor 2A) and suppresses general protein translation. ERN1 interacts with TRAF2 (TNF receptor-associated factor 2) and activates the kinases IKBK (inhibitor of kappa light polypeptide gene enhancer in B cells) and MAPK8. This activates the transcription factors NFκB and AP1 and increases the expression of inflammatory cytokines. In metabolic tissues, such as WAT and liver, the activation of MAPK8 also leads *via* serine phosphorylation of IRS1 to defective insulin actions, such as insulin resistance (Sect. 9.2).

ERN1 has also endoribonuclease activity and cleaves, *e.g.*, the *XBP1* mRNA encoding for the transcription factor X-box binding protein 1, which results in the translation of an activated form of XBP1 responsible for upregulation of many chaperone genes. Furthermore, ATF6 translocates from the ER to the Golgi apparatus, where it is processed by proteases to an active transcription factor (ATF6f). Finally, ER stress also leads to the cleavage and activation of the transcription factor CREB3L3 (cAMP responsive element binding protein 3-like 3), which induces, particularly in the liver, the production of the acute-phase protein CRP. The goal of activating the three arms of the unfolded protein response is to restore ER homeostasis by

- reducing general protein synthesis
- facilitating protein degradation
- increasing the protein folding capacity (Fig. 7.6b).

The lipophilic environment of the large ER membrane is provided with important functions in the metabolism of lipids, in particular of phospholipids and cholesterol. For example, cholesterol sensing is initiated at the ER membrane through SREBF1 (Sect. 3.1). This indicates a direct connection between lipid metabolism and the unfolded protein response, such as the control of ER phosphatidylcholine synthesis and ER membrane expansion by XBP1. Moreover, ER stress is linked to the production of inflammatory mediators, such as the enzyme PTGS2 (prostaglandin-endoperoxide synthase 2, also known as COX2) and ROS. This disturbs lipid metabolism and glucose homeostasis leading to abnormal insulin action, promotes hyperglycemia through insulin resistance, stimulates hepatic glucose production and suppresses glucose disposal. When the unfolding protein response cannot reconstitute proper ER function or when the metabolic stress continues, apoptotic pathways are initiated, *i.e.*, the affected cells are dying. This happens, *e.g.*, to foam cells during atherosclerosis. Thus, **nutrient and inflammatory responses are integrated in metabolic homeostasis, but dysfunction of the ER affects this integration and results in chronic metabolic disease** (Chap. 10). For example, the reciprocal regulation between ER stress and insulin signaling pathways leads to a vicious cycle explaining the interdependence of insulin resistance (Sect. 9.2) and atherosclerosis (Sect. 10.2).

Additional Readings

Franceschi C, Garagnani P, Parini P, Giuliani C, Santoro A (2018) Inflammaging: a new immune-metabolic viewpoint for age-related diseases. Nat Rev. Endocrinol 14:576–590

Galluzzi L, Yamazaki T, Kroemer G (2018) Linking cellular stress responses to systemic homeo-stasis. Nat Rev. Mol Cell Biol 19:731–745

Tall AR, Yvan-Charvet L (2015) Cholesterol, inflammation and innate immunity. Nat Rev. Immunol 15:104–116

Wang A, Luan HH, Medzhitov R (2019) An evolutionary perspective on immunometabolism. Science 363

Zmora N, Bashiardes S, Levy M, Elinav E (2017) The role of the immune system in metabolic health and disease. Cell Metab 25:506–521

Chapter 8
Obesity

Abstract In this chapter, we will define obesity as the consequence of excess WAT accumulation that increases the risk of non-communicable diseases. We will describe adipocytes as the central cellular component of adipose tissue and adipogenesis as the key process creating fat cells. In response to appropriate signals, such as low temperature, white adipocytes are able to transform to beige adipocytes displaying a phenotype similar to brown adipocytes. During the development of overweight and obesity, adipocytes first grow in size and then in number attracting many M1-type macrophages. The latter form together with T cells the major stromal-vascular fraction of adipose tissue and can lead to chronic inflammation in the tissue. We will show that adipokines have a major impact during hypertrophy and hyperplasia of WAT and for the communication with the CNS. Studying monogenic forms of obesity provides strong evidence for a central role of appetite regulation in obesity susceptibility. The leptin-melanocortin pathway has an integral role in this satiety signaling. Variations in genes of this pathway as well as numerous others will be presented as important drivers of common obesity in context of the modern obesogenic environment.

Keywords Obesity · BMI · Visceral fat · Subcutaneous fat · White, beige and brown adipocytes · Adipogenesis · Chronic inflammation · Adipokines · Leptin · Hindbrain · Hypothalamus · MC4R · FTO

8.1 Definition of Obesity

No other tissue of our body can change its dimension as dramatically as adipose tissue. This is accomplished first by increasing the size of individual cells up to a critical threshold (**hypertrophy**) and then increasing the number by recruiting new adipocytes from the resident pool of progenitors (**hyperplasia**). The *WHO* defines overweight and obesity as "abnormal or excessive fat accumulation that may impair health". **Obesity is the consequence of excess WAT growth and develops when energy intake exceeds energy expenditure**. The most commonly used measure of obesity is the BMI (Sect. 1.4). A person is defined as normal weight if his/her BMI is 18.5–24.9 kg/m^2, overweight if the BMI is 25–29.9 or obese if the BMI >30. Individuals with adult-onset of obesity mostly exhibit increased adipocyte size, whereas persons with early-onset obesity show both adipocyte hypertrophy and hyperplasia. Thus, the number of adipocytes in a given fat depot is determined early in life and is mostly stable through adulthood. However, differentiated adipocytes have remarkable hypertrophic potential, since they are able to increase in size to several hundred μm in diameter. Moreover, the location of WAT in the body plays an important role for the risk to develop the metabolic syndrome (Chap. 10). High amount of visceral fat (distributed in the abdominal cavity), referred to as central or **"apple-shaped" obesity**, increases the risk, while rise in subcutaneous fat (localized beneath the skin), referred to as peripheral or **"pear-shaped" obesity**, exerts far less risk (Fig. 8.1a).

Obesity is rare to occur in the wild life, but there are examples of animals living in harsh climates, such as polar bears and seals, that are obese. This indicates that a high degree of natural obesity can even contribute to evolutionary fitness. However, in humans obesity mostly occurs with low-grade chronic inflammation (Sect. 8.3) and consecutively is often accompanied by the different features of the metabolic syndrome (Chap. 10). In fact, most of today's world population lives in countries where individuals are more likely to die from the consequences of being obese than starving (Box 8.1). Of note, human obesity does not always result in disease suggesting that the threshold for tolerable BMI differs among individuals and may be determined by environmental and genetic variables (Sect. 8.5).

8.2   Adipogenesis

Adipose tissue is not only a passive storage container for nutrients but also an active endocrine organ that communicates with our body. This communication is mediated by nutritional mechanisms, neural pathways (Sect. 8.4) and autocrine, paracrine and endocrine actions of secreted proteins that are collectively referred to as adipokines (Table 8.1). Since most adipokines act pro-inflammatory and only a few anti-inflammatory, their overall expression is increased in the obese state compared to the lean state. **The adipokine secretion profile of adipocytes significantly changes during the onset of obesity**. Pro-inflammatory adipokines are the peptide hor-

---

Fig. 8.1 (continued) body can also be measured using anthropometric measures, such as waist circumference or waist-to-hip ratio (WHR). Obese subjects with a low WHR, characterized as pear-shaped obesity with predominantly increased subcutaneous fat, have a lower risk of T2D and metabolic syndrome. In contrast, obese subjects with a high WHR, characterized as apple-shaped obesity with increased visceral fat, have a high risk of for these diseases. WAT is found in all areas of our body (**b**). The subcutaneous and the intraabdominal depots are the main fat storage compartments. BAT is abundant at birth and is still present in adults, but to a lesser extent

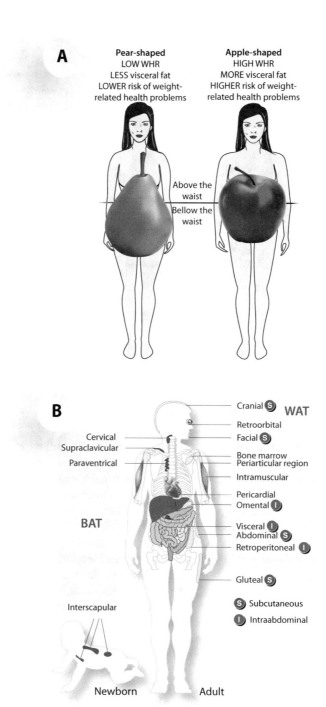

Fig. 8.1 Fat distribution influences obesity-associated risks. Obesity is defined by a BMI of ≥30 and in general is a consequence of fat accumulation (**a**). The respective fat distribution in the

Box 8.1: Worldwide Increase of Overweight and Obesity

The prevalence of obesity has increased worldwide in the past 40 years and reached pandemic levels. Between 1980 and 2013, the proportion of overweight or obese adults (BMI >25) increased worldwide from 28.8% to 36.9% in males and from 29.8% to 38.0% in females. Obesity substantially increases the risk of many non-communicable diseases, such as T2D, fatty liver disease, hypertension, myocardial infarction, stroke, dementia, osteoarthritis, obstructive sleep apnoea and several cancers (Sects. 1.4 and 1.5). Thus, **obesity contributes to a decline in both quality of life as well as of life expectancy**. Worldwide prevalence of obesity increased from 1975 to 2016 in children and adolescents from 0.7% to 5.6% in boys and 0.9% to 7.8% in girls. Within the same time period, the worldwide prevalence of obesity increased from 3.2% to 10.8% in adult men and from 6.4% to 14.9% in adult women. The obesity prevalence in adults varies by country and ranges from 3.7% in Japan to 38.2% in the United States. Since 2006 the dynamics in adult obesity has slowed down in developed countries, while the prevalence of obesity exceeded 50% in men in Tonga and in women in Kuwait, Kiribati, Micronesia, Libya, Qatar, Tonga and Samoa. Nevertheless, **at all ages prevalence of overweight and obesity was higher in developed than in developing countries**. During the last three decades no single country showed a significant decrease in obesity implying the danger that over time most countries are on a trajectory to reach the same high rates as already observed in Tonga or Kuwait.

mones leptin and resistin, the transport proteins RBP4 (retinol binding protein 4) and lipocalin 2, the growth factors ANGPTL2 (angiopoietin-like protein 2), the enzyme NAMPT, the cytokines TNF, IL6 and IL18 and the chemokine CXCL5 (C-X-C motif ligand 5). In contrast, only the adipokines adiponectin and SFRP5 (secreted frizzled-related protein 5) act anti-inflammatory. The balance between pro-inflammatory and anti-inflammatory adipokines is crucial for determining homeostasis throughout the body based on the nutritional status (Box 8.2).

Thus, **adipose tissue influences and communicates *via* adipokines with many other organs**, such as the brain, heart, liver and skeletal muscle, and vascularization, respectively. During adipose tissue expansion adipocyte dysfunction often occurs, such as a dysregulation of adipokine production, which has both local and systemic effects on inflammatory responses. This significantly contributes to the initiation and progression of obesity-induced cardiovascular and metabolic diseases (Chap. 10).

Storing excess fat in vesicles is an evolutionary adapting process, which already started in worms, such as *C. elegans*. Most vertebrate species store fat in a tissue of a mesodermal origin, named WAT. Also in humans WAT is the large majority of adipose tissue, *i.e.*, it is the primary site of energy storage. In contrast, minor amounts of fat are BAT (brown adipose tissue), which is a site of basal and inducible

Table 8.1 Sources and functions of key adipokines

Adipokine	Primary source(s)	Binding partner or receptor	Function
Adiponectin	Adipocytes	Adiponectin receptors 1 and 2, T-cadherin, calreticulin-CD91	Insulin sensitizer, antiinflammatory
SFRP5	Adipocytes	WNT5A	Suppression of pro-inflammatory, WNT signaling
Leptin	Adipocytes	Leptin receptor	Appetite control through the central nervous system
Resistin	PBMCs	Unknown	Promotes insulin resistance and inflammation through IL6 and TNF secretion from macrophages
RBP4	Liver, adipocytes, macrophages	Retinol (vitamin A), transthyretin	Implicated in systemic insulin resistance
Lipocalin 2	Adipocytes, macrophages	Unknown	Promotes insulin resistance and inflammation through TNF secretion from adipocytes
ANGPTL2	Adipocytes, other cells	Unknown	Local and vascular inflammation
TNF	Stromal vascular fraction cells, adipocytes	TNF receptor	Inflammation, antagonism of insulin signaling
IL6	Adipocytes, stromal vascular fraction cells, liver, muscle	IL6 receptor	Changes with source and target tissue
IL18	Stromal vascular fraction cells	IL18 receptor	IL18 binding protein, broad-spectrum inflammation
CCL2	Adipocytes, stromal vascular fraction cells	CCR2	Monocyte recruitment
CXCL5	Stromal vascular fraction cells (macrophages)	CXCR2	Antagonism of insulin signaling through the JAK-STAT pathway
NAMPT	Adipocytes, macrophages, other cells	Unknown	Monocyte chemotactic activity

---

Box 8.2: Adipokines

Leptin is the most important hormone produced by adipose tissue, since it regulates feeding behavior through the CNS (Sect. 8.4). Moreover, leptin has effects on cells of the immune system and stimulates the production of pro-inflammatory cytokines and chemokines in monocytes and macrophages, such as TNF, IL6 and CXCL5. Furthermore, leptin polarizes T cells towards a T_H1 phenotype. The peptide hormone **resistin** is also associated with inflam-

(continued)

Box 8.2 (continued)

mation, since it promotes the expression of the pro-inflammatory cytokines TNF and IL6. Resistin induces insulin resistance *via* SOCS3, which is an inhibitor of insulin signaling (Sect. 6.3). Moreover, resistin directly counteracts the anti-inflammatory effects of adiponectin on vascular endothelial cells. The main source of **RBP4** is the liver, but also adipocytes and macrophages can produce the transporter of vitamin A (retinol). In an auto- or paracrine manner RBP4 inhibits insulin-induced phosphorylation of IRS1, *i.e.*, the adipokine is involved in the regulation of glucose homeostasis in adipocytes. **Lipocalin 2** belongs to the same protein superfamily as RBP4 and transports various small lipophilic substances, such as retinoids, arachidonic acid and steroids. The protein is induced by inflammatory stimuli through the activation of NFκB. High lipocalin 2 concentrations are found in obese individuals. **ANGPTL2** is a growth factor that induces inflammatory responses and activates integrin signaling in endothelial cells, monocytes and macrophages. It can induce insulin resistance and its serum levels are associated with obesity, insulin resistance and CRP concentrations. The enzyme **NAMPT** (also called visfatin) is mainly expressed and secreted by adipose tissues. NAMPT is essential for the biosynthesis of NAD and has an important role in controlling the insulin secretion of β cells (Sect. 3.6). **TNF** is a pro-inflammatory cytokine with a prominent role in basically all inflammatory and autoimmune diseases. The cytokine is mainly produced by monocytes and macrophages, but it can also be secreted by activated adipocytes. TNF promotes insulin resistance in skeletal muscle and adipose tissues by reducing the phosphorylation of IR and IRS1 (Sect. 6.3). TNF concentrations are increased in adipose tissue and plasma of obese individuals. **IL6** is also a pro-inflammatory cytokine that is involved in obesity-related insulin resistance. Adipose tissue is a major source of the cytokine, since more than 30% of all circulating IL6 is produced there. Another pro-inflammatory cytokine produced by adipose tissues is **IL18**. Atherosclerotic lesions show high IL18 levels and indicate plaque instability. The chemokine **CXCL5** is secreted by macrophages within the stromal vascular fraction of adipose tissue and is associated with inflammation and insulin resistance. CXCL5 interferes with insulin signaling in muscles by activating the JAK (Janus kinase)-STAT (signal transducer and activator of transcription) pathway through its receptor CXCR2 (CXC-chemokine receptor 2). The peptide hormone **adiponectin** is exclusively synthesized by adipocytes and is found at high levels in serum. Adiponectin is expressed at the highest levels in functional adipocytes of lean persons, while its expression is downregulated in dysfunctional adipocytes of obese individuals. The beneficial effects of adiponectin on insulin sensitivity are mediated *via* increased Ca^{2+} levels in skeletal muscle that activate CAMKK2, AMPK and SIRT1 and result in the upregulation of *PPARGC1A* expression (Sect.

(continued)

Box 8.2 (continued)
6.6). The main function of the anti-inflammatory adipokine **SFRP5** is to prevent the binding of WNT (wingless-type MMTV integration site family member) proteins to their respective receptors. The WNT signaling pathway has a number of important downstream targets, such as MAPK8, leading to pro-inflammatory cytokine production in macrophages.

energy expenditure (Fig. 8.1b). WAT is found throughout our body, such as around the omentum, intestines and perirenal areas, as well as subcutaneously in the buttocks, thighs and abdomen. Moreover, WAT arises also on the face and extremities and within the bone marrow. In newborns, BAT occurs in the neck, kidneys and adrenal regions, while in adults it locates in the neck as well as in supraclavicular and paravertebral regions.

WAT buffers nutrient availability and demand by storing excess calories and preventing toxic lipid levels in non-adipose tissues. This is an essential function for survival, because it allows intervals of fasting between meals and intervals of prolonged fasting. BAT maintains core body temperature in response to cold stress by generating heat, *i.e.*, it is primarily used for non-shivering thermogenesis. White adipocytes have one large lipid droplet filling 90% of the cell, while brown adipocytes carry many single lipid compartments and a far larger number of mitochondria than WAT. These mitochondria are enriched with the long-chain fatty acid/H^+ symporter UCP1, which causes a proton leak across the inner mitochondrial membrane, *i.e.*, it uncouples fuel oxidation from ATP synthesis. When white adipocytes express high amounts of UCP1, they turn into "beige" adipocytes (Fig. 8.2), a process referred to as "browning". In reverse, these cells can again increase lipid storage and then morphologically resemble classic white adipocytes, referred to as "whitening". **This tissue conversion is an adaptive process, *i.e.*, it depends on environmental challenges, such as low temperatures for browning or a high-fat diet for whitening**.

Mesenchymal stem cells are the precursors to fat, bone and muscle cells. Growth factors, such as BMPs (bone morphogenetic proteins) and FGFs, are central in the first phase of adipogenesis. In this commitment phase, the multipotent mesenchymal stem cells differentiate to WAT and BAT precursors. Both brown adipocytes and myocytes derive from paraxial mesoderm-derived progenitor cells that express the transcription factors MYF5 (myogenic factor 5) and PAX7 (paired box 7) (Fig. 8.2). However, in adults brown adipocytes can also develop from skeletal muscle satellite cells. White adipocytes derive from both $MYF5^-$ and $MYF5^+$ progenitors. In the second phase of adipogenesis, the differentiation phase, transcription factors, such as the nuclear receptor PPARγ (Sect. 3.3) and the pioneer factors CEBPA, CEBPB and CEBPD, are both necessary and sufficient for adipogenesis. Moreover, co-factors of these transcription factors, such as PPARGC1A (Sect. 6.2),

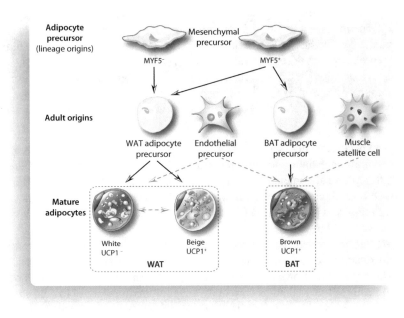

Fig. 8.2 Origins of white, beige and brown adipocytes. BAT contains UCP1 expressing brown adipocytes (UCP1$^+$), whereas WAT is formed by UCP1$^-$ white adipocytes and UCP1$^+$ beige adipocytes. In adults, the expansion of adipose tissue is mainly achieved through the growth and differentiation of preadipocytes (*i.e.*, adipocyte precursors). The precursors of WAT and BAT adipocytes origin from mesenchymal cells: for WAT they derive from both MYF5$^+$ and MYF5$^-$ lineages, whereas for BAT they come exclusively from the MYF5$^+$ lineage. Beige adipocytes are obtained from WAT adipocyte precursors or directly from mature white adipocytes. In contrast, brown adipocytes can derive from stem cell-like skeletal muscle satellite cells. In addition, brown and white adipocytes are generated from endothelial precursors

support adipogenesis. Furthermore, chromatin modifiers, such as the lysine methyltransferase EHMT1 or the deacetylase SIRT1 (Sect. 6.6), control in adipocytes the access of chromatin for transcription factors and their co-factors, *i.e.*, **adipogenesis involves epigenome-wide changes**.

White, beige and brown adipocytes can undergo adaptive and dynamic changes in response to starvation or overfeeding as well as in response to cold environment *via* energy-sensing pathways. One central and illustrative example is the transformation of white adipocytes into beige adipocytes at cold temperature exposure (Fig. 8.3). Some of the signals that regulate this tissue conversion are synthesized locally within the adipose tissue, *i.e.*, they act paracrine. However, other essential factors are of endocrine nature, as they are produced by metabolic organs, such as brain, muscle, heart and liver. In response to exposure to cold temperature, catecholamines, such as adrenaline, are released by the SNS (sympathetic nervous system). Interestingly, in response to cold stress catecholamines are also secreted by M2-type

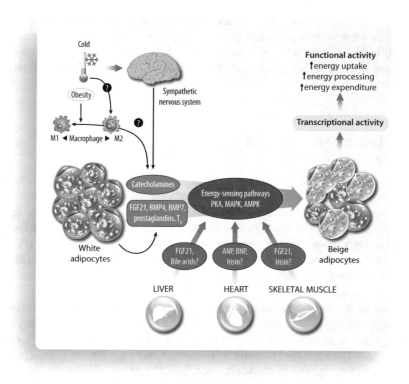

Fig. 8.3 Hormonal control of WAT browning. Metabolic adaptions to environmental factors are regulated by the release of endocrine and paracrine factors from metabolic tissues. In response to (thermal) cold, catecholamines are released by the SNS and from M2-type macrophages in adipose tissue. This activates energy-sensing pathways in white adipocytes and stimulates their transformation to beige adipocytes. This beige phenotype is generated through the actions of transcription factors that induce activities characteristic of their phenotype, such as increase energy uptake, energy processing and energy expenditure

macrophages in adipose tissue. This is counteracted by M1-type macrophages that are present in hypertrophic adipose tissue of obese individuals (Sect. 8.3).

Catecholamine-activated β3-adrenergic receptors, PTGS2-generated prostaglandins and growth factors are the key molecules that promote browning of white adipocytes (Fig. 8.3). BMP4 and BMP7 directly regulate thermogenesis in mature brown adipocytes by increasing their responsiveness to catecholamines and upregulating intercellular lipase activity *via* the PKA-MAPK signal transduction pathway. The transformation of white into beige adipocytes is further supported by the growth factors FGF21 and BDNF (brain-derived neurotrophic factor) as well as the peptide hormone irisin, which is primarily produced by skeletal muscle in response to exercise. Interestingly, the amount of BAT in our body correlates with the month when we were conceived. Individuals conceived in cold months have significant differences in BAT characteristics and metabolic phenotypes than those, who were conceived in warm months or environment. An analogous mouse experiment confirmed the observation in rodents and demonstrated that only the cold exposure of the fathers before conception affected the amount of brown adipose tissue in the offspring.

8.3 Inflammation in Adipose Tissue

Adipose tissue is a metabolic organ that is formed by parenchymal cells and stromal cells. In WAT, lipid-laden adipocytes represent only 20–40% of the cell number but more than 90% of its volume. Every gram of adipose tissue contains 1–two million adipocytes but 4–six million stromal-vascular cells, of which **more than half are immune cells, such as macrophages and T cells**. In the healthy state (Fig. 8.4, stage 1), these cell types work together, in order to maintain metabolic homeostasis. Also in disease these tissues try to interact, in order to adapt to altered conditions, such as increased nutritional needs of the affected organs.

Adipose tissue can be classified into at least three structural and functional groups. Lean individuals with normal metabolic function store excess nutrients as triacylglycerols in WAT. The WAT in these lean subjects contains M2-type macrophages and T_H2 cells that respond to nutrient-derived signals by promoting lipid storage and suppressing lipolysis (Fig. 8.4, stage 1). When obesity develops as the result of chronic overnutrition, the storage capacity is exceeded causing cellular dysfunction, such as lipid dysregulation, mitochondrial dysfunction, oxidative stress and ER stress (Sect. 7.4), leading to reduced metabolic control. This causes adipocytes to secrete chemokines, such as CCL2, that attract monocytes into the adipose tissue that become M1-type macrophages (Fig. 8.4, stage 2). Adipocytes increase in size until they reach a structurally critical condition, in which the vascularization of the tissue is reduced, so that adipocytes experience hypoxic conditions. When these alterations escalate, they lead to adipocyte death. In order to remove remnants of dead adipocytes, additional macrophages infiltrate the WAT. They surround the dead cells and create crown-like structures that are associated with increased inflammation (Fig. 8.4, stage 3).

Stromal cells, such as macrophages, support adipocytes in WAT in their main metabolic function, the long-term storage of lipids. The number and activation state of macrophage both reflect the metabolic health of WAT. In lean persons, only 10–15% of the stromal cells are macrophages and most of them are of M2-type. These M2-type macrophages secrete IL10 that potentiates insulin action in adipocytes, *i.e.*, it maintains or even increases the insulin sensitivity of these cells. During the development of obesity, WAT recruits monocytes that differentiate into M1-type macrophages and finally can comprise up to 60% of all stromal cells in the tissue. These M1-type macrophages are a major driver of insulin resistance in WAT, but

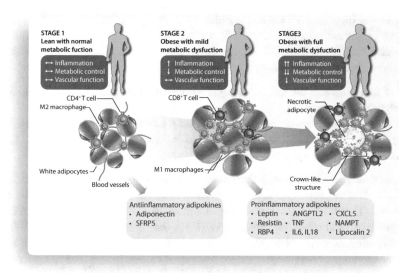

Fig. 8.4 Functional classification of adipose tissue. Adipose tissue can be distinguished into at least three stages. In normal-weight tissue with normal metabolic function (stage 1) adipocytes are associated with a rather low number of M2-type macrophages. This tissue produces preferentially anti-inflammatory cytokines, such as adiponectin and SFRP5 (Sect. 8.2). During onset of obesity, adipocytes increase their triglyceride storage, *i.e.*, they become hypertrophic. At limited obesity (stage 2) adipocytes still retain relatively normal metabolic function and display low levels of immune cell activation and sufficient vascular function. However, in obesity with full metabolic dysfunction (stage 3) the tissue has recruited a large number of M1-type macrophages and produces preferentially pro-inflammatory adipokines, such as leptin, resistin, RBP4, lipocalin, ANGPTL2, NAMPT, TNF, IL6, IL18 and CXCL5

they are also involved in the remodeling of the enlarging adipocytes. Thus, **the two types of macrophages coordinate homeostatic adaptations of adipocytes in the lean and the obese state**.

T cells in adipose tissue also play a role in obesity-induced inflammation. T_H1 cells produce pro-inflammatory cytokines, such as IFNγ, while T_H2 cells and T_{REG} cells secrete anti-inflammatory cytokines, such as IL10, inducing differentiation of macrophages into M2 type. In lean individuals, T_H2 and T_{REG} cells dominate in WAT, while in obese persons there are far more T_H1 cells. Compared to subcutaneous fat, visceral fat accumulates a larger number of macrophages and secretes greater amounts of pro-inflammatory cytokines. In addition, adipocytes in visceral fat are more fragile and reach earlier a critical size triggering cell death than subcutaneous adipocytes. This explains, at least in part, the different health risk between the apple and pear shapes of fat depots (Sect. 8.1).

The long-term exposure of adipocytes with pro-inflammatory cytokines produced by M1-type macrophages can induce insulin resistance of WAT (Sect. 9.2).

Like in acute microbe infection, this reduced insulin sensitivity initially tries to react to the increased levels of nutrients by limiting their storage. However, the strategy of inducing insulin resistance becomes maladaptive in case of a constant, long-term nutrient overload. Thus, the hallmarks of obesity-induced inflammation, also referred to as "metaflammation", are that it

- is a nutrient-induced inflammatory response orchestrated by WAT-associated macrophages
- changes the polarization of these macrophages from M2 to M1 phenotype
- represents a moderate/low-grade and local expression of inflammatory cytokines is chronic without apparent resolution.

In other metabolic organs, polarized macrophages play a comparable role. In the BAT, resident macrophages differentiate into M2-type after exposure to cold temperatures. These M2-type macrophages induce thermogenic genes in BAT and lipolysis of stored triacylglycerols in WAT *via* the secretion of the catecholamine noradrenaline. Kupffer cells, the resident macrophages of liver, enable the metabolic adaptations of hepatocytes during increased caloric intake. The M2 phenotype of Kupffer cells is induced *via* PPARδ and the T_H2-type cytokines IL4 and IL13. Under the condition of obesity M2-type macrophages regulate fatty acid β-oxidation in the liver and support in this way hepatic lipid homeostasis. Similar to WAT, in the pancreas high-fat diet induces the infiltration of M1-type macrophages. The increased intake of dietary lipids results in β cell dysfunction, which induces the expression of chemokines recruiting inflammatory macrophages to the islets. The secretion of IL1B and TNF by the infiltrating macrophages further augments β cell dysfunction (Sect. 9.5).

8.4   Energy Homeostasis and Hormonal Regulation of Food Uptake

Energy homeostasis in adults is achieved by a combination of processes that manage energy intake, energy storage in form of glycogen in liver, kidney and skeletal muscles and triacylglycerols in WAT and energy usage, in order to maintain a stable body weight. Thus, food intake is an integrated response over a prolonged period of time that maintains the levels of energy stored in adipocytes. However, **as the result of a daily tiny but cumulative positive energy balance, overweight and obesity can develop in the course of many years**. This misbalanced energy homeostasis is multifactorial and complex (Fig. 8.5). Food consumption has changed radically the last generation(s), leading to dramatic changes in our macronutrient intake (Sect. 1.1). This is based on reduced physical activity, computer-based work dominating most occupations, leisure time entertainment becoming dependent on information technology, reduction in home cooking, greater reliance on convenience food, a growing habit of snack consumption and promotion of large portions. Thus, **today's obesogenic environment is the most likely cause of the obesity epidemic** (Sect. 8.1).

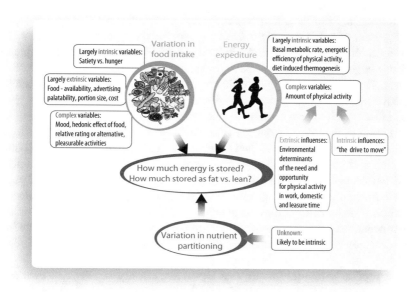

Fig. 8.5 Variables of energy homeostasis. Energy homeostasis is regulated by a complex interaction between the variation in food intake, the tendency to store excess energy as fat or lean mass, referred to as nutrient partitioning, and the variation in energy expenditure. The intrinsic variables that have an impact on energy homeostasis causing obesity are influenced by food intake through effects on satiety and hunger

The feelings of hunger and satiety are the main involuntary motivations for feeding-related behavior of humans and animals. The coordinated secretion of numerous hormones from the CNS prepares the digestive system for the anticipated caloric load. Ideally, satiation hormones (being secreted in response to ingested nutrients) control the amount of food intake. In turn, obesity hormones (indicating the fat content of the body) modify these signals. However, many non-homeostatic factors, such as stress, cultural habits and social influences interact with these hormonal controllers of food intake. In theory, establishing a negative energy balance, *i.e.*, eating less calories than daily consumed by the basal metabolic rate and additional physical activity, would be an easy solution in reducing the problem of overweight and preventing the development of obesity. However, **hormonal and neuronal control circuits have been trained by evolutionary adaption to make hunger the first ranking desire of nearly all humans, which strongly counteracts to most attempts losing body weight**. In addition, changes in environmental exposures early in life, before, during or after pregnancy, *i.e.*, nutrition-triggered epigenetic programming (Sect. 5.4), cause a sustained long-term effect on the predisposition to develop overweight and obesity. In general, genetic variants that cause severe familial obesity largely influence food intake through effects on hunger and satiety (Sect. 8.5). Accordingly, **variations associated with obesity are predominantly in genes expressed in the brain**.

Energy homeostasis is largely based on the coordinated activity of multiple peptide hormones. Prior an anticipated meal, the gastrointestinal tract is prepared for the digestion of nutrients and for avoiding extreme metabolic consequences of the pending caloric load. For example, individuals who habitually eat at the same time each day begin these CNS-initiated hormone secretions, such as insulin, before food serving. This is important for the efficient disposal of absorbed glucose (Sect. 9.1). Moreover, CNS signals also stimulate secretion of the peptide hormone ghrelin from the stomach approximately 30 min before the meal and the incretine GLP1 (glucagon-like peptide 1) from the intestine rises even 1 h earlier. When food is consumed, numerous hormones and enzymes are secreted, in order to allow nutrient digestion and absorption. **Most of the hormones related to digestion, such as CCK (cholecystokinin), are satiety signals**. GLP1, glucagon, APOA4 and peptide YY are further gastrointestinal peptides that influence the hindbrain to reduce the meal size. The half-life of these peptides is short and the function of most of them is redundant, *i.e.*, they can compensate each other. Satiation signals converge in the NTS (nucleus tractor solitarius) and the adjacent area postrema of the hindbrain (Fig. 8.6).

The peptide hormones **insulin and leptin are obesity signals**, *i.e.*, their secretion is proportional to the amount of body fat. Together with ghrelin, they enter the brain through the blood-brain barrier *via* receptor-mediated active transport and act directly on the ARC (arcuate nucleus) of the hypothalamus (Fig. 8.6). This area of the brain also receives information from the hindbrain on the progress of meals and from limbic centers reflecting non-homeostatic influences. In addition, the peptide hormone amylin that is secreted like insulin from β cells of the pancreas, as well as cytokines derived from adipose tissue, such as IL6 and TNF, all act on the brain, in order to increase energy expenditure.

The ARC contains two populations of neurons that either express NPY (neuropeptide Y) and AGRP or POMC (proopiomelanocortin) (Fig. 8.6). POMC is a prohormone that is cleaved to produce α-MSH (α-melanocyte-stimulating hormone) being a ligand of MC4R (melanocortin 4 receptor). In contrast, insulin and leptin activate POMC neurons that release α-MSH counteracting AGRP and stimulating MC4R neurons. This results in reduced food intake and decreased body weight in the long-term inhibition of food intake. Ghrelin stimulates NPY-AGRP neurons that secrete AGRP acting as an antagonist of MC4R neurons. This leads to increased food intake and induction of weight gain. These areas of the forebrain also receive information from the hindbrain concerning the progress of the meal and information from non-homeostatic factors. Reduced leptin and insulin signaling in the ARC results in increased food intake and eventually weight gain. In contrast, the cytokine TNF, derived from macrophages within adipose tissue, reduces food intake. Insulin and leptin can also regulate ongoing food intake by directly adjusting the sensitivity to incoming satiation signals in the NTS of the hindbrain. For example, when an individual loses weight, the secretion of insulin and leptin decreases and the reduced obesity signal in the brain results in reduced sensitivity to the satiating action of CCK and GLP1. This leads to an increase in meal size until lost weight is regained and the levels of obesity signals are restored. The opposite can be observed when

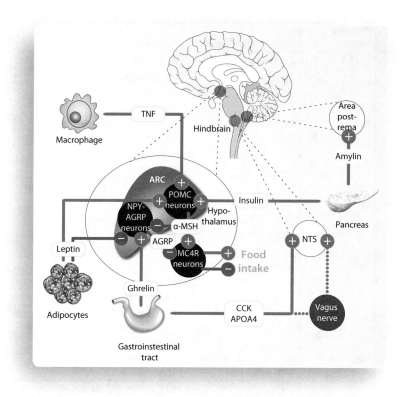

Fig. 8.6 Hormonal signals from the periphery influence multiple brain areas. Insulin and amylin from the pancreas stimulate POMC neurons in the ARC of the hypothalamus and in the area postrema of the hindbrain, respectively. Ghrelin stimulates NPY-AGRP neurons in the ARC, while leptin from adipocytes inhibits these cells. However, leptin and TNF stimulate POMC neurons. Gastrointestinal peptides, such as CCK and APOA4, either stimulate the NTS directly or stimulate vagal afferent nerves whose axons end in the NTS. Most of these peptide hormones enter the brain *via* active transport

people overeat and gain weight. Thus, **the homeostatic control of food intake is integrated within the brain by an intermingled network of hypothalamic and brainstem structures**.

8.5 Genetics of Obesity

Some rare forms of severe obesity result from mutations in an individual gene or chromosomal region, *i.e.*, they represent monogenic obesity (Table 8.2). The importance of the leptin-melanocortin pathway in hyperphagia (*i.e.*, increased appetite)

Table 8.2 Monogenic cases of obesity

Gene	Genomic position	Mode of inheritance	Associated phenotype
POMC	2p23.3	Autosomal recessive	Severe pediatric-onset obesity
			Hyperphagia
			Red hair pigmentation
			Pale skin
LEP	7q32.1	Autosomal recessive	Severe early-onset obesity
			Extreme hyperphagia
			Hyperinsulinemia
			Hypothalamic hypothyroidism
			Hypogonadotropic hypogonadism
LEPR	1p31.3	Autosomal recessive	Severe obesity with hyperphagia
			Delayed or absent puberty
			Reduced IGF1 levels
			Growth abnormalities
PCSK1	5q15	Autosomal recessive	Severe childhood obesity
			Abnormal glucose homeostasis
			Reduced plasma insulin with elevated proinsulin levels
			Hypogonadotropic hypogonadism
			Hypocortisolemia
MC4R	18q21.32	Autosomal dominant/ recessive	Severe early-onset obesity
			Hyperphagia
			Highly elevated plasma insulin concentrations
			Increased bone mineral density
SIM1	6q16.3	Autosomal dominant	Early-onset obesity
			Hypotonia
			Developmental delay
			Short extremities

and obesity susceptibility is indicated by the fact that so far primarily mutations in the genes *LEP* (leptin), *LEPR*, *POMC*, *PCSK1* (proprotein convertase subtilisin/kexin type 1), *MC4R* and *SIM1* (single-minded family bHLH transcription factor 1) were found as causes of monogenetic obesity. *PCSK1* encodes for an enzyme responsible for post-translational processing of POMC, while *SIM1* encodes for a transcription factor that is both an upstream and downstream target of MC4R. Accordingly, **MC4R mutations are responsible for up to 6% of childhood obesity and 2% of adult obesity cases**. Importantly, the very rare obesity phenotype of patients with homozygous *LEP* mutations can be reversed by administration of leptin.

The rapid increase in the number of obese people (Sect. 8.1) can be explained by radical changes in lifestyle, such as high intake of energy-dense food and physical inactivity (Sect. 8.3). However, some subjects are more susceptible to these lifestyle changes than others, suggesting a relevant genetic component. Polygenic common obesity results from the combined effect of multiple genetic variants in concert with environmental risk factors. Linkage analysis, candidate gene approaches and in particular GWAS (Sect. 2.4) in various populations have indicated dozens of genes to be associated with the traits BMI and obesity. Widely replicated candidate genes are *MC4R*, *BDNF*, *PCSK1*, *ADRB3* (adrenoceptor beta 3) and *PPARG*. The most prominent result from GWAS analysis was the identification of a strong association of the chromosomal region of the *FTO* gene with BMI and obesity. However, not the *FTO* gene but its neighboring genes *IRX3* and *IRX5* encoding for transcription factors functionally explain the prominent effect on obesity risk (Box 4.1) Although the effect size of the genetic variations at the *FTO* locus is not comparable to that of monogenic forms, it represents the most established association with common obesity, primarily due to its high frequency (47%) in the European population.

GWAS meta-analysis of nearly 340,000 individuals identified 97 genomic loci associated with BMI, confirming the central role of the CNS, in particular of genes expressed in the hypothalamus, in the regulation of body mass. Thus, **obesity can be considered as a neurobehavioral disorder with high susceptibility to an obesogenic environment**. However, the known risk genes in total account only for 2.7% of the variation of the trait. This observation questions, whether genetic variations are the key cause of obesity. Thus, a substantial portion of the predicted heritability of obesity and interindividual variability in BMI remains unexplained. This implies that the understanding of obesity needs to be extended by epigenetic and social-behavioral components.

Additional Readings

Afshin A, Sur PJ, Fay KA, Cornaby L, Ferrara G, Salama JS, Mullany EC, Abate KH, Abbafati C, Abebe Z et al (2019) Health effects of dietary risks in 195 countries, 1990–2017: a systematic analysis for the Global Burden of Disease Study 2017. Lancet 393:1958–1972

Blüher M (2019) Obesity: global epidemiology and pathogenesis. Nat Rev Endocrinol 15:288–298

Challet E (2019) The circadian regulation of food intake. Nat Rev Endocrinol 15:393–405

Fetissov SO (2017) Role of the gut microbiota in host appetite control: bacterial growth to animal feeding behaviour. Nat Rev Endocrinol 13:11–25

Ghaben AL, Scherer PE (2019) Adipogenesis and metabolic health. Nat Rev Mol Cell Biol 20:242–258

Ng M, Fleming T, Robinson M, Thomson B, Graetz N, Margono C, Mullany EC, Biryukov S, Abbafati C, Abera SF et al (2014) Global, regional, and national prevalence of overweight and obesity in children and adults during 1980–2013: a systematic analysis for the Global Burden of Disease Study 2013. Lancet 384:766–781

Chapter 9
Insulin Resistance and Diabetes

Abstract In this chapter, we will describe the molecular principles of glucose homeostasis and insulin signaling as well as their dysregulation leading to insulin resistance and β cell failure. Since our blood glucose levels need to stay within a physiological range of 4–6 mM, glucose intake, storage, mobilization and breakdown are tightly regulated. Insulin plays a key role in these regulatory processes. When normal concentrations of insulin cause an insufficient response of the major insulin target tissues, such as skeletal muscle, liver and adipose tissue, insulin resistance has developed. Ectopic overload of lipids, chronic inflammatory response and ER stress are the main processes that can lead to insulin resistance. Moreover, glucotoxic and lipotoxic stress to β cells of the pancreas are mediated *via* inflammatory response, oxidative stress and ER stress eventually resulting in the failure of the cells, *i.e.*, in the inability to produce insulin. We will describe diabetes as a disease of dysregulation of glucose and lipid homeostasis that does not only affect the insulin production in pancreatic β cells but also the metabolism in organs, such as liver, muscle and fat. Worldwide, the prevalence of T2D is rapidly increasing, which, when not properly treated, ultimately leads to reduced life expectancy due to microvascular (retinopathy, nephropathy and neuropathy) and macrovascular (heart disease and stroke) complications. Both genetic and environmental factors contribute to the development of the disease. We will realize that despite large GWAS screening for risk genes less than 10% of the inheritance of T2D is understood. Therefore, epigenome-wide changes, both prenatal as well as in adult life, play an important role in the disease.

Keywords Glucose homeostasis · Insulin resistance · Chronic inflammation · Ectopic lipid deposition · Glucotoxicity · Lipotoxicity · ER stress · Oxidative stress · Unfolded protein response · β cell failure · T1D · T2D · OGTT · Insulin · β cells · Liver · Skeletal muscle · Adipose tissue · Inflammation · MODY · GWAS · Epigenetic programming

© Springer Nature Switzerland AG 2020   131
C. Carlberg et al., *Nutrigenomics: How Science Works*,
https://doi.org/10.1007/978-3-030-36948-4_9

9.1   Glucose Homeostasis

Glucose homeostasis results both from the hormonal and neural control of glucose production and use, which even at physiological challenges, such as food ingestion, fasting and intense physical activity, maintains the blood glucose level within a range of 4–6 mM. This constant level is essential for providing energy to tissues, most importantly for an uninterrupted glucose supply to the brain and red blood cells, which almost exclusively use glucose as an energy source. Hypoglycemia, *i.e.*, a blood glucose level below 4 mM, can lead in the brain to a number of neuro-glycopenic effects. In contrast, a constant concentration above 10 mM, *i.e.*, **chronic hyperglycemia, causes glucotoxicity in blood vessels leading to a number of complications in the cardiovascular system, the kidneys, the eyes and nerves** (Box 9.1).

Box 9.1: Complications of Chronic Hyperglycemia
People with diabetes (both T1D and T2D) are at risk of developing a number of troubling, disabling and life-threatening health problems. Chronically high blood glucose levels, *i.e.*, glucotoxicity, can lead to serious diseases affecting the blood vessels of brain, heart, eyes, kidneys and peripheral nerves. In almost all high-income countries, diabetes is a leading cause of CVD, blindness, kidney failure and lower-limb amputations.

CVDs: This is the most common cause of disability and death among people with diabetes. CVDs that accompany diabetes include cerebral stroke, myocardial ischemia, congestive heart failure and peripheral artery disease (Chap. 10).

Kidney insufficiency: Nephropathy finally resulting in kidney failure is caused by damage to small blood vessels, through which the kidneys function less efficiently or even fail. Diabetes is one of the leading causes of chronic kidney disease.

Eye disease: In retinopathy the network of blood vessels, which supply the retina becomes blocked and damaged. In addition, pathological neovascularization leads to increasingly loss of vision, finally to complete blindness.

Nerve damage: In neuropathy nerves throughout the body are damaged, which can lead to problems with digestion and micturition (a reflex producing a series of contractions of the urinary bladder), erectile dysfunction and a number of other dysfunctions. The most commonly affected areas are the extremities, particularly the lower legs and feet (peripheral neuropathy) leading to pain, paresthesia (abnormal dermal sensation) and loss of feeling. The latter is particularly dangerous because even small injuries will be unnoticed, leading to ulcerations, superinfections and finally to major amputations (diabetic foot syndrome).

The principal regulators of glucose homeostasis are the peptide hormones glucagon and insulin that are secreted by α and β cells, respectively, of the endocrine pancreas (forming Langerhans islets). Glucagon is secreted when blood glucose concentration is low, such as between meals and during exercise. Glucagon has the greatest effect on the liver, where it stimulates the release of glucose that was stored in form of glycogen into the blood and the production of glucose *via* the gluconeogenesis pathway.

In contrast, rising blood glucose levels directly after food ingestion stimulates insulin secretion. The glucose transporter GLUT2 in the plasma membrane of β cells and the hexokinase GCK in the cytoplasma both sense glucose (Sect. 3.1) and initiate glucose import and its metabolism *via* glycolysis (Fig. 9.1). The increasing ATP-ADP ratio stimulates the closing of ATP-sensitive K^+ (K^{ATP}) channels, plasma membrane depolarization, activation of voltage-gated Ca^{2+} channels and Ca^{2+}-mediated stimulation of exocytosis of insulin granules. This K^{ATP} channel-dependent mechanism is a triggering signal that is particularly important for the acute phase of insulin release, *i.e.*, during the first 10 min after glucose rise. In the second phase of insulin secretion, the mitochondrial glucose metabolism generates signals additional to the ATP-ADP ratio, which are important for gaining insight into the functional failure of β cells during T2D. Pyruvate, the final product of glycolysis, flows into mitochondria through an anaplerotic (*i.e.*, "refilling") process *via* the enzyme PC (pyruvate carboxylase) and an oxidative pathway *via* the PDH (pyruvate dehydrogenase) complex (Fig. 9.1). The conversion of pyruvate to oxaloacetate *via* PC and the following metabolism of oxaloacetate to malate, citrate or isocitrate in the TCA cycle provides several possibilities for the reconversion of these metabolites into pyruvate *via* cytosolic and mitochondrial pathways. **These metabolic pathways are important for the regulation of glucose-stimulated insulin secretion.** One of these pathways is the export of citrate from the mitochondria through the citrate-isocitrate carrier SLC25A1 and the subsequent conversion of isocitrate to α-ketoglutarate by the cytosolic NADP-dependent IDH (isocitrate dehydrogenase) complex. The metabolic byproducts of this pyruvate-isocitrate cycling may also act as amplifying signals for the control of glucose-stimulated insulin secretion.

After food ingestion, carbohydrates are digested in the gastrointestinal tract and glucose is absorbed into the circulation primarily *via* the hepatic portal vein. The liver has a central role in monitoring and regulating post-prandial (*i.e.*, after a meal) glucose levels (Fig. 9.2, left). Insulin promotes hepatic synthesis of triacylglycerol and their storage in WAT upon feeding. Moreover, insulin also suppresses the release of stored lipids from adipose tissue. Ingested nutrients in intestinal endocrine cells stimulate the release of incretins, such as GLP1, that together with the rise in blood glucose stimulate β cells to deliver insulin. **The first phase of insulin secretion primarily prevents the liver from producing more glucose by stimulating glycogen synthesis and suppressing gluconeogenesis. The second phase, approximately 1–2 h after the meal, stimulates glucose uptake by insulin-sensitive tissues, such as skeletal muscle and adipose tissue.**

During fasting (Fig. 9.2, right), hepatic glycogenolysis decreases as hepatic glycogen stores deplete. Low insulin levels combined with elevated counter-regulatory

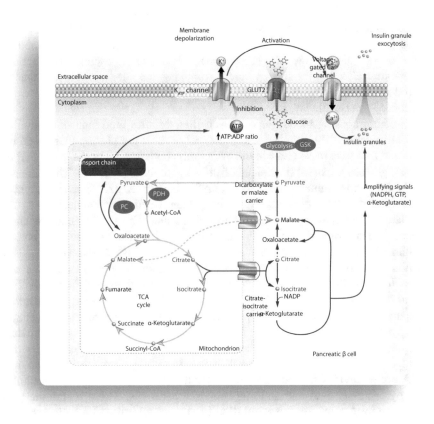

Fig. 9.1 Glucose-stimulated insulin secretion in β cells. Rising blood glucose levels stimulate GLUT2 in the membrane of β cells to import glucose and GCK to start glucose breakdown *via* glycolysis generating ATP. The increased ATP-ADP ratio inhibits K^{ATP} channels resulting in membrane depolarization, activation of voltage-gated Ca^{2+} channels, influx of Ca^{2+} and stimulation of insulin granule exocytosis. Moreover, pyruvate is the end product of glycolysis and enters mitochondrial metabolism *via* PDH or PC. β cells also exhibit active "pyruvate cycling" *via* the anaplerotic entry of pyruvate or other substrates into the TCA cycle generating excess of intermediates that then exit the mitochondria to engage in various cytosolic pathways leading back to pyruvate. Pyruvate-isocitrate cycling generates an amplifying signal that enhances the Ca^{2+}-mediated triggering signal for insulin exocytosis

hormones, such as glucagon, adrenaline and corticosteroids, promote hepatic glucose production *via* gluconeogenesis, so that **blood glucose levels remain stable across a wide range of physiological conditions, such as fasting**. Moreover, glucagon and adrenaline stimulate lipolysis in WAT and fatty acid β-oxidation in other tissues when nutrients are in limited supply. Although the hormonal regulation of glucose homeostasis is essential, also the CNS can sense and respond to acute changes in glucose and nutrient needs through innervation of the intestine, the liver, the pancreas, the portal vein and all other glucose-demanding tissues (Sect. 8.4).

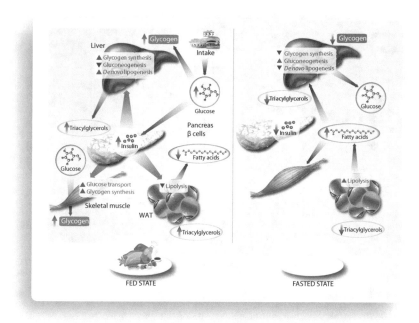

Fig. 9.2 Insulin action in health. In the fed state, dietary carbohydrates increase plasma glucose levels and promote insulin secretion from β cells. In skeletal muscle, insulin increases the transport of glucose and permits glucose entry and glycogen synthesis. In adipose tissue, insulin suppresses lipolysis and promotes *de novo* lipogenesis. In the liver, insulin stimulates glycogen synthesis and *de novo* lipogenesis and inhibits gluconeogenesis. In the fasted state, insulin secretion is decreased, which increases hepatic gluconeogenesis and promotes glycogenolysis. Under these conditions, hepatic lipid production diminishes while adipose lipolysis increases

9.2   Insulin Resistance in Skeletal Muscle and Liver

Insulin resistance is a condition, in which normal concentrations of insulin produce a subnormal biological response in insulin target tissues. β cells compensate this reduced response by increasing the production of insulin. As long as this hyperinsulinemia is adequate to overcome the insulin resistance, glucose tolerance remains relatively normal. In patients destined to develop T2D, the β cell compensatory response fails (Sect. 9.5) and insulin insufficiency develops leading to impaired glucose tolerance and eventually T2D. **Impaired insulin sensitivity causes impaired insulin-stimulated glucose uptake into skeletal muscle, impaired insulin-mediated inhibition of hepatic glucose production in liver and a reduced ability of insulin to inhibit lipolysis in WAT**.

There are three main mechanisms to explain insulin resistance, in particular in muscle cells (Fig. 9.3a) and liver cells (Fig. 9.3b). These are

- an ectopic (*i.e.*, unusual) lipid accumulation in muscle and liver

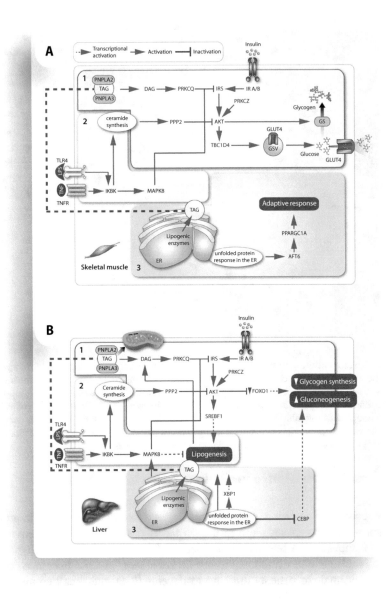

Fig. 9.3 Pathways involved in skeletal muscle (a) and hepatic (b) insulin resistance. The insulin-IR-IRS-PI3K-AKT signaling axis promotes *via* TBC1D4 the translocation of GSVs (GLUT4-containing storage vesicles) to the plasma membrane permitting the entry of glucose into the cell and also stimulates glycogen synthesis *via* the enzyme GS. This central signal transduction pathway is connected to a number of additional pathways. Green shaded (1): DAG-mediated activation of the kinase PRKCQ and the subsequent inhibition of IRS, ceramide-mediated increases in the AKT inhibitor PPP2 and increased sequestration of AKT by the kinase PRKCZ. Impaired

- a chronic inflammatory response of the tissues
- ER stress mediated *via* the unfolding protein response.

All three mechanisms are originally derived from an evolutionary advantage in adapting to a changing environment. The evolutionary oldest pathway, the unfolded protein response (Sect. 7.5), was designed to integrate metabolic signals, such as metabolic stress through lipid overload, and to adapt accordingly. The inflammatory response (Sect. 7.2) also represents an evolutionary old pathway that is highly interconnected with the unfolded protein response, in order to provide a coordinated response to various environmental stimuli, such as nutrient scarcity. This requires the dampening of the insulin response, in order to allow a metabolic shift from glucose to lipid oxidation.

Humans that perform long-term fasting shift into an insulin resistant mode in one or several tissues, such as muscle, liver or WAT, in order to preserve blood glucose for the brain. This insulin resistance leads to increased fatty acid concentrations in the circulation as well as in skeletal muscle and liver. The elevated lipid levels in these tissues also cause a parallel increase in lipids with signaling function, such as DAG (diacylglycerol) and ceramides. This then enhances the insulin resistance of the tissues and ensures the preservation of glucose for the CNS. However, **this natural mechanism for survival became pathogenic in the modern times, where often the level of energy intake exceeds the level of energy expenditure, *i.e.*, in conditions of misbalanced energy homeostasis** (Sect. 8.4).

The lipid content in muscle cells reflects a net balance between fatty acid uptake and their oxidation in mitochondria. Thus, **acquired mitochondrial dysfunction is an important predisposing factor for ectopic lipid accumulation and insulin resistance in the elderly**. LPL (Sects. 3.4 and 7.3) is a key enzyme for the hydrolysis of circulating triacylglycerols (*e.g.*, within VLDLs) that permits the uptake of fatty acids in muscle and liver through a complex of fatty acid transport proteins of the SLC27A family with the scavenger receptor CD36 (Fig. 9.3). Upon entry into the cell, fatty acids are rapidly esterified to acyl-CoAs. These are successively transferred to a glycerol backbone to form mono, di-, and triacylglycerols or esterify with sphingosine to form ceramides. This means that the levels of the second messengers DAG and ceramide rise in parallel to the increased lipid load of the cells. Intracellular lipid droplets carry on their surface enzymes, such as the lipases PNPLA (patatin-like phospholipase domain containing) 2 and PNPLA3, that regulate the entry and exit of lipid molecules and catalyze their lysis, *e.g.*, from triacylglycerols to DAG. Thus, **PNPLAs are essential both for the access of the energy stored in triacylglycerols and the generation of lipid mediators of insulin resistance**.

Fig. 9.3 (continued) AKT2 activation limits translocation of GSVs to the plasma membrane, resulting in impaired glucose uptake. It also decreases insulin-mediated glycogen synthesis. Yellow shaded (2): Inflammatory pathways, such as the activation of IKBK affecting ceramide synthesis and the activation of MAPK8 inhibiting IRS *via* phosphorylation. Pink shaded (3): The unfolded protein response in the ER leading to activation of ATF6 and a PPARGC1A-mediated response. Key lipogenic enzymes in the ER membranes stimulate lipid droplet formation. TAG = triacylglycerol

Both in muscle and in liver DAG activates members of the PRKC family, such as PRKCQ, that impair insulin signaling *via* inhibition of IRS1 and IRS2 resulting in a decreased glucose uptake *via* GLUT4 (Fig. 9.3a). Ceramides dampen insulin signaling by the activation of PPP2 (protein phosphatase 2) dephosphorylating AKT and *via* PRKCZ that binds AKT and prevents its activation. When in the liver the rates of DAG synthesis from fatty acid re-esterification and *de novo* lipogenesis exceed the rates of lipid oxidation in the mitochondria, *i.e.*, when there is an increase of DAG levels, PRKCE is activated, IR tyrosine kinase activity is inhibited, GSK3 is hyperphosphorylated and glycogen synthesis is decreased. Furthermore, this leads to increased translocation of FOXO to the nucleus promoting elevated expression of gluconeogenic enzymes (Fig. 9.3b). Inflammatory mediators and adipokines, such as TNF, secreted from adipose tissue, can act locally in a paracrine manner or they leak out of the adipose tissue causing a systemic effect (endocrine action) on insulin sensitivity in muscle and liver cells. *Via* the TNFR (TNF receptor) signaling axis this activates MAPK8 and IKBK. Thus, **the inflammatory response results in the inactivation of IRS1 and leads to insulin resistance**.

Stress of the ER caused by the accumulation of unfolded proteins in its lumen plays a special role in the pathogenesis of insulin resistance in the liver (Fig. 9.3). Activation of three key proteins of the unfolded protein response, EIF2AK3, ERN1 and ATF6, results in increased membrane biogenesis, stop of protein translation and elevated expression of chaperone proteins in the ER. *Via* the activation of MAPK8 this leads to inhibitory serine phosphorylation of IRS1. Moreover, the unfolded protein response results in an expansion of the ER membrane and increases the expression of SREBF1 (Sect. 3.1), which stimulates lipogenesis. Thus, **the unfolded protein response causes hepatic insulin resistance, when it is able to alter the balance of lipogenesis and lipid export to promote hepatic lipid accumulation**.

9.3   β Cell Failure

The failure of β cells during the progression to T2D involves their chronic exposure to glucose and lipids, also known as glucotoxicity and lipotoxicity, or in combination "glucolipotoxicity". A chronic glucose exposure increases glucose metabolism in β cells leading to the formation of citrate, which acts as a signal for the formation of malonyl-CoA in the cytosol (Fig. 9.4a). Malonyl-CoA inhibits the key fatty acid transporter in mitochondria, CPT1A (carnitine palmitoyltransferase

----

Fig. 9.4 (continued) protein misfolding. The protein unfolding response is initially able to balance this ER stress, but over time this becomes less effective, and the deleterious effects of ER stress leads to cell death. Insulin hypersecretion is accompanied by amylin secretion forming amyloid fibrils that accumulate at the surface of β cells and induce dysfunction and apoptotic death to the cells. Prolonged hyperglycemia results in oxidative, ER and hypoxic stress and in increased exposure of β cells with cytokines (**b**). Therefore, β cells may cease their proliferation, de-differentiate or undergo uncontrolled autophagy or apoptosis. All these processes reduce the number of β cells and their function. This leads dysfunction and depletion of β cells and to progress of T2D

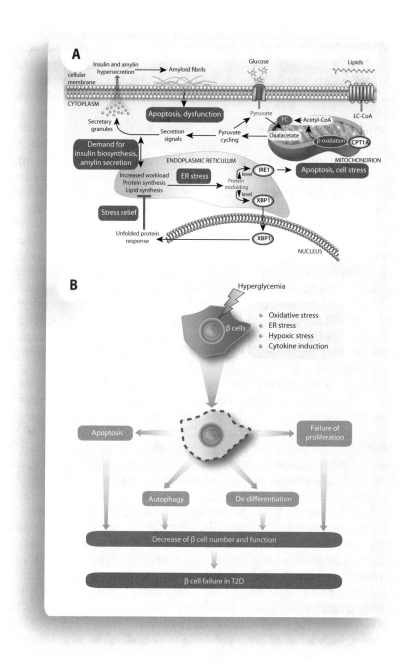

Fig. 9.4 β cell failure. Overnutrition and/or increased lipid supply induces in mitochondria of β cells enzymes of fatty acid β-oxidation, such as CPT1A, resulting in increased acetyl-CoA levels, allosteric activation of PC and constitutive upregulation of pyruvate cycling (**a**). This leads to increased basal secretion of insulin and a loss of the glucose-stimulated increment in the flux of pyruvate cycling, *i.e.*, blunting of glucose-stimulated insulin secretion. The elevated demand for synthesis of insulin in the ER increases the stress to this organelle, resulting in elevated rates of

1A) and blocks in this way fatty acid β-oxidation. This causes accumulation of SFA-CoAs in β cells. The high demand for insulin secretion during hyperglycemia creates significant metabolic stress to the ER of β cells and results in the overproduction of ROS in their mitochondria. This oxidative stress is a central element of glucotoxicity.

When intracellular glucose concentrations exceed the glycolytic capacity of β cells, some of the molecules are converted to enediol intermediates, leading to superoxide formation. Since β cells contain only low levels of anti-oxidant enzymes, such as catalase, glutathione peroxidase and superoxide dismutase 2, they are very susceptible to superoxide damage. Moreover, β cells have many mitochondria and consume more oxygen than most other cell types. Therefore, at high glucose conditions, such as directly after a meal, the accelerated mitochondrial function enhances the oxygen consumption and causes hypoxia. This parallels with the increased expression of hypoxia-inducible genes. In addition, T2D patients have an expanded ER in their β cells indicating increased stress to the organelle. One of the stress sensors is the ER transmembrane protein IRE1 (inositol-requiring enzyme), which can induce apoptosis. Moreover, the hyperinsulinemia in response to chronic hyperglycemia disrupts ER homeostasis in β cells due to the exceeded capacity for proinsulin biosynthesis. This leads to accumulation of misfolded proteins and induction of the unfolded protein response *via* XBP1 (Sect. 7.5), which enhances the oxidative stress and eventually results in β cell dysfunction.

Patients with a long clinical history of T2D commonly have a decreased β cell number and function, respectively, which is often referred to as "β cell exhaustion". Compared with weight-matched healthy individuals, obese T2D patients have a 63% reduction of β cell mass, while lean T2D patients only show a 41% loss. This suggests that β cell dysfunction has a primary role in the pathogenesis of T2D. In response to ER stress, hypoxic stress and exposure to pro-inflammatory cytokines, β cells fail to proliferate or undergo apoptosis or uncontrolled autophagy (Fig. 9.4b). Moreover, the β cells can de-differentiate or trans-differentiate into other pancreatic cell types, such as α cells. β cells proliferate *via* the replication of preexisting β cells and the differentiation of progenitor cells, referred to as neogenesis. In the pancreas of adults, the dominant mechanism for increasing β cell numbers is replication rather than neogenesis. The mass of β cells is controlled by the balance between the rate of proliferation and the rate of apoptosis. The FAS (Fas cell surface death receptor) pathway, a central apoptosis regulatory mechanism, is upregulated in patients with poorly controlled T2D. Thus, **both increased apoptosis or decreased proliferation can reduce the β cell mass of T2D patients**.

Moreover, high levels of saturated FFAs, *i.e.*, lipotoxicity, also induce β cell apoptosis. Interestingly, in patients who are genetically predisposed to T2D, but not in healthy individuals, a sustained increase in plasma FFA levels causes β cell dysfunction.

9.4   Definition of Diabetes

Diabetes is a condition of chronically elevated plasma glucose levels, referred to as hyperglycemia, that eventually causes toxicity to blood vessels (Box 9.1). There are two major forms of diabetes, T1D and T2D. T1D results from an autoimmune destruction of insulin-producing β cells in the pancreas. As a result, the body can no longer produce insulin and the respective patients need insulin injections every day for the rest of their life, in order to control the levels of glucose in their blood. Without insulin, a person with T1D will die premature. This type of diabetes has a sudden onset and usually affects children of 10 years or older, *i.e.*, in an age when their immune system has reached full potency. **The number of people who develop T1D is increasing, which may be due to changes in environmental risk factors, prenatal events, diet early in life or viral infections**.

T2D is the most common type of diabetes, representing more than 90% of all diabetes cases. It usually occurs in adults, but is increasingly seen in children and adolescents. In initial stages of T2D, β cells are still able to produce insulin, but either the amounts are insufficient or the body is unable to respond to its effects (known as insulin resistance, Sect. 9.2), both leading to elevated glucose levels in the blood. T2D often remains unnoticed and undiagnosed for years, *i.e.*, the respective persons are unaware of the already smouldering long-term damage being caused by their disease. In contrast to people with T1D, the majority of T2D patients usually do not require daily doses of insulin to survive. Many T2D patients are able to manage their hyperglycemia through a healthy diet and increased physical activity or by oral medication for a rather long time (Sect. 9.5). However, when they reach a stage, in which they are unable to regulate their blood glucose levels, they need insulin substitution.

Women, who during pregnancy (mostly around the 24th week) develop a resistance to insulin and subsequent high blood glucose levels, have gestational diabetes (17% of live births to women in 2013). **Uncontrolled gestational diabetes can have serious consequences for both the mother and her baby and increases the risk of the child to develop T2D later in life**.

An OGTT (oral glucose tolerance test) with measurements of glucose at defined times (*e.g.*, at 0, 30, 60 and 120 min) after oral uptake of a defined amount of glucose (often 75 g) is the easiest way to determine the glucose homeostasis status of prediabetic individuals. Healthy persons have a fasting blood glucose level in the order of 5 mM, already 1 h after the glucose bolus show a peak below 10 mM and return to less than 7.8 mM after 2 h (Fig. 9.5, No. 1). Individuals that start at normal glucose concentrations but after 2 h still have levels higher than 7.8 mM have impaired glucose tolerance (Fig. 9.5, No. 2). However, when the fasting glucose level exceeds 7 mM and after 2 h still is higher than 11.1 mM, the person is considered diabetic (Fig. 9.5, No. 3). The response measured in the OGTT reflects the ability of β cells to secrete insulin and the responsiveness of the whole body to

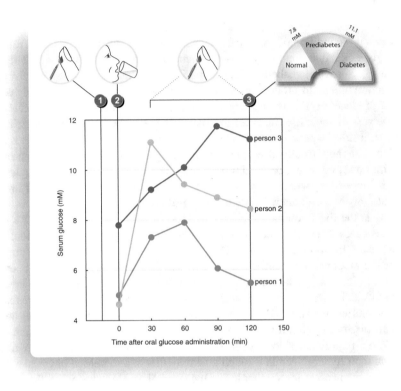

Fig. 9.5 Oral glucose tolerance test. The test measures how the human body responds to an oral challenge of glucose (usually as a drink of 75 g). Blood glucose is measured in a time course (*e.g.*, every 30 min over 2 h). The glucose level increases quickly, but the secretion of insulin should manage the normalization of the glucose concentration after 2 h (5 mM, person No. 1). Person No. 2 has normal fasting plasma glucose levels, but due to impaired glucose tolerance does not return after 2 h to normal concentrations (below 7.8 mM). In contrast, person No. 3 is diabetic, since his/her fasting glucose level already exceeds 7.8 mM, and the 2 h value is clearly elevated, respectively

insulin. For example, a person with a fasting glucose in the range of 6.1–7.0 mM is categorized to have impaired fasting glucose and may have established insulin resistance (Sect. 9.2). **These individuals have impaired glucose homeostasis and are at increased risk to develop T2D.**

The worldwide T2D prevalence of adults is 8.5% (2017) and this rate will further increase (Fig. 9.6). The incidence of diabetes rises when countries become more industrialized, people eat a more sugar- and fat-rich diet and are less physical active. Despite the predominantly urban impact of the T2D epidemic, is rapidly becoming also a major health concern in rural communities in low- and middle-income countries. In high-income countries, primarily people above the age of 50 years get T2D, while in middle-income countries the highest prevalence is in younger persons. As these populations age, the prevalence will rise further due to the increase of older age groups. **The mortality rate of diabetes varies sharply with the economy of the country being significantly lower in high-income countries with a more developed healthcare system**.

9.5   Failure of Glucose Homeostasis in T2D and Its Treatment

T2D is an age-related disease that is strongly promoted by overnutrition and physical inactivity. In the early stages of T2D, insulin levels rise to maintain glucose tolerance by compensating for increased insulin resistance of skeletal muscles and adipose tissue (Sect. 9.2). While the insulin resistance of an individual remains relatively constant over time, the deterioration of the insulin-secretory capacity of β cells increases continuously. Under these conditions, insulin is less potent in suppressing hepatic glucose production, *i.e.*, also the liver becomes insulin resistant. In later stages of the disease, β cells get exhausted and lose their ability to compensate

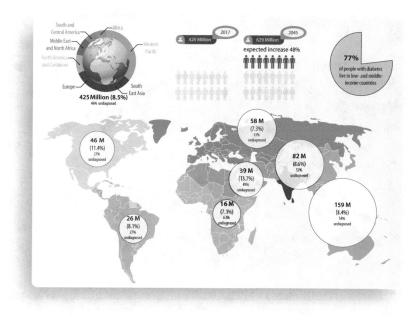

Fig. 9.6   T2D in numbers. The majority of the 425 million people with T2D (2017) are between 40 and 59 years old. The worldwide prevalence of the disease is 8.5%. Until 2045, the number of people with diabetes will increase by 48%. No country escapes the T2D epidemic. Data were obtained from https://diabetesatlas.org

via the increase of insulin release. This results in reduced circulating insulin concentrations and often occurs in parallel with increased glucagon levels. The shift in the glucagon/insulin ratio leads to a further rise in hepatic gluconeogenesis, *i.e.*, the liver releases more glucose to the circulation. When basal in addition to postprandial blood glucose levels are chronically increased, the individual develops hyperglycemia. Moreover, defective insulin signaling also causes dyslipidemia (Sect. 10.3), including perturbed homeostasis of fatty acids, triacylglycerols and lipoproteins (Fig. 9.7).

The defective insulin secretion and responses in T2D have several reasons. Firstly, constant exposure of β cells to elevated levels of glucose and lipids, *i.e.*, glucolipotoxicity, induces their dysfunction and ultimately triggers their death (Sect. 9.3). These processes are related to chronic inflammation of pancreatic islets. Elevated glucose levels increase the metabolic activity of the islet cells, in which *via* increased ROS production, the NLRP3 inflammasome is activated (Sect. 7.1). Secondly, increased insulin demand and production induces ER stress to β cells

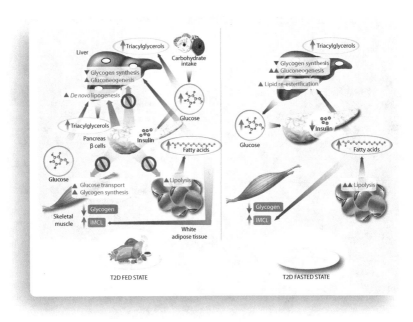

Fig. 9.7 Insulin actions in diabetes. In T2D, insulin-mediated skeletal muscle glucose uptake is impaired, which directs glucose to the liver. Increased hepatic lipid levels impair the ability of insulin to regulate gluconeogenesis and to stimulate glycogen synthesis. However, lipogenesis in liver is not affected. In combination with increased delivery of dietary glucose, this stimulates lipogenesis causing NAFLD. Impaired insulin action in adipose tissue increases lipolysis, which promotes re-esterification of lipids in other tissues, such as the liver, and further exacerbates insulin resistance. In combination with a decline in the number of active β cells, this leads to the development of hyperglycemia. *IMCL* intramyocellular lipid

(Sect. 7.4) further activating the inflammasome. Like in other inflammatory scenarios, this cytokine production leads to the attraction of macrophages and other immune cells. Furthermore, the islets produce an amyloid polypeptide that aggregates to form amyloid fibrils in patients with T2D. Thus, **resident islet macrophages adopt a pro-inflammatory M1 phenotype that induces islet dysfunction**.

Current treatments for T2D include insulin, the indirect AMPK activator metformin, K^{ATP} channel inhibiting sulphonylureas, PPARγ-activating thiazolidinediones (glitazones), incretin mimetics and their degradation-inhibitors, and inhibitors of either starch- and disaccharides-digesting α-glucosidase or of glucose transporters. Each of these therapies can improve hyperglycemia and some may even delay the onset of diabetes. However, none of these drugs can slow down the progressive decline in insulin secretion. Intensive diabetes treatment results in tight glycemic control and therefore a substantial reduction in the risk of microvascular complications (Box 9.1). Since T2D is often associated with hypertension (Sect. 10.1) and dyslipidemia (Sect. 10.3), respective drugs are prescribed to most patients with T2D in addition to glucose-lowering medications. Importantly, **T2D can be prevented by lifestyle changes**. For example, already a moderately increase in physical activity combined with a decrease in caloric intake, aiming for a persistent 5–10% weight loss, reduces the risk for T2D by more than 50%.

9.6 Genetics and Epigenetics of T2D

A range of monogenetic disorders result in chronic hyperglycemia (Table 9.1). They are summarized as MODY (maturity onset diabetes of the young), because they often occur already in young adults. However, the therapy of these inherited forms of diabetes does not require insulin, *i.e.*, they are of non-T1D type. Most of the MODY genes encode for transcription factors, such as *HNF4A*, *HNF1A*, *HNF1B*, *PDX1*, *NEUROD1* (neuronal differentiation 1), *KLF11* (Krüppel-like factor 11) and *PAX4*. In contrast *GCK*, *CEL* (carboxyl ester lipase) and *BLK* (B lymphoid tyrosine kinase) encode for enzymes, *ABCC8* and *KCNJ11* (potassium inwardly-rectifying channel, subfamily J, member 11) for ion channels, *APPL1* (adaptor protein, phosphotyrosine interacting with PH domain and leucine zipper 1) for an adaptor protein and *INS* (insulin) for a hormone.

Monogenetic forms of T2D represent only 1–2% of all diabetes cases worldwide. In contrast, typical obesity-related T2D is often found to carry a cluster of genetic variations that confer enhanced susceptibility to environmental factors, such as overnutrition and stress. All MODY genes are expressed in β cells and affect insulin secretion, while the normal control of glucose metabolism *via* insulin involves a number of additional organs, such as muscle, liver and fat (Sect. 9.1). This suggests that **insulin secretion in β cells is a more important parameter for diabetes than insulin resistance in peripheral organs**.

Table 9.1 MODY genes

Type	OMIM	Gene/ protein	Description
MODY 1	125850	*HNF4A*	Due to a loss-of-function mutation in the *HNF4A* gene. 5–10% of cases.
MODY 2	125851	*GCK*	Due to any of several mutations in the *GCK* gene. 30–70% of cases. Mild fasting hyperglycemia throughout life. Small rise on glucose loading.
MODY 3	600496	*HNF1A*	Mutations of the *HNF1A* gene. 30–70% of cases. Tend to be responsive to sulfonylureas. Low renal threshold for glucose.
MODY 4	606392	*PDX1*	Mutations of the *PDX1* gene. Less than 1% of cases. Associated with pancreatic agenesis in homozygotes and occasionally in heterozygotes.
MODY 5	137920	*HNF1B*	Defect in *HNF1B* gene. 5–10% of cases. Atrophy of the pancreas and several forms of renal disease
MODY 6	606394	*NEUROD1*	Mutations of the *NEUROD1* gene. Very rare.
MODY 7	610508	*KLF11*	Mutations of the *KLF11* gene.
MODY 8	609812	*CEL*	Mutations of the *CEL* gene. Very rare. Associated with exocrine pancreatic dysfunction.
MODY 9	612225	*PAX4*	Mutations of the *PAX4* gene.
MODY 10	613370	*INS*	Mutations in the *INS* gene. Usually associated with neonatal diabetes. Less than 1% of cases.
MODY 11	613375	*BLK*	Mutated *BLK* gene. Very rare.
MODY 12	606391	*ABCC8*	Mutated *ABCC8* gene. Very rare.
MODY 13	616329	*KCNJ11*	Mutated *KCNJ11* gene. Very rare.
MODY 14	616511	APPL1	Mutated *APPL1* gene. Very rare.

80% of cases of early-onset, autosomal-dominant, familial hyperglycemia are represented by the 14 MODY genes

T2D belongs to those diseases that have been extensively studied by GWAS. Figure 10.4 provides an overview on 18 genetic variants that were the first to be associated with T2D. They all represent common SNPs with MAFs ranging from 7.3% to 50%. The gene *TCF7L2* (transcription factor 7-like 2) has an OR of 1.37 for developing T2D (*i.e.*, a 37% increased T2D risk), while the ORs for the 17 remaining genes range only between 1.05 and 1.15 (5–15% increased risk). These numbers are comparable to what is observed as genetic risk for other common traits and diseases, such as obesity (Sect. 8.5). Some of the T2D risk genes, such as *CDKAL1* (CDK5 regulatory subunit associated protein 1-like 1), *SLC30A8* encoding for a zinc transporter, *HHEX* (hematopoietically expressed homeobox) encoding for a transcription factor, and *KCNJ11*, are involved in insulin secretion in β cells. Thus, **the common risk for T2D agrees with the findings of monogenetic diabetes**.

However, the genetic propensity to develop T2D involves also genes in a number of additional pathways that affect β cell formation and function, as well as pathways affecting fasting glucose levels and obesity. The cluster of *CDKN* (cyclin-dependent kinase inhibitor) *2A* and *CDKN2B* controls β cell growth, *MTNR1B* (melatonin receptor 1B) links circadian rhythms with fasting glucose levels (Sect. 3.6), *FTO/IRX3/IRX5* is the key risk gene locus for obesity (Sect. 8.5), *PPARG* encodes for the master regulator of adipogenesis (Sect. 8.2) and *IGF2BP2* (insulin-like growth factor 2 mRNA binding protein 2) is involved in insulin signaling (Sect. 6.3). However, for the eight remaining T2D risk genes no direct link to metabolic homeostasis had been identified indicating that not always the closest genes to the T2D-associated SNP provide a functional explanation but in some cases more distant genes may be involved (Fig. 9.8). The 18 genetic variants explain only some 4% of the heritable risk for T2D. Nevertheless, in personalized medicine approaches, such as exemplified by iPOP (Sect. 4.6), most prominent T2D risk genes are already used as predictive markers. Larger study populations and meta-analyses of existing studies increased the number of T2D risk gene loci to more than 100 (www.genome. gov/gwastudies). The additional genes have similar or even lower MAFs and ORs. Thus, from the genetic perspective **T2D is a very heterogenous disease that can be segregated into multiple subtypes, which should be treated on a personalized basis taking the individual's genetic background and phenotype into account**.

In general, common SNPs are characterized by low ORs, while rare monogenetic forms of T2D have high ORs (Fig. 9.9). However, both extremes do not explain all genetic basis of T2D. Like for many other common diseases and traits, also for T2D risk there is a large number of low frequency SNPs with intermediate ORs. Some of these genetic variants are expected be identified by the use of whole genome sequencing (Sect. 2.5), but will not be able to explain all heritability of T2D. Thus, **prenatal and post-natal epigenetic programming will demonstrate its contribution to the disease risk**.

The experience from the already discussed *Dutch hunger winter* (Sect. 5.3) provides a molecular explanation for an increased obesity and T2D risk. Similar observations for a rise in T2D prevalence had been made for survivors of the *Ukraine famine* (1932–1933) and the *Chinese famine* (1959–1961). The *in utero* environment has a strong impact on fetal epigenetic programming, as individuals exposed to either famine or maternal gestational diabetes during fetal development develop more likely obesity and/or T2D later in life (Fig. 9.10a). **Food-deprived conditions of the mother change the epigenome of the fetus, so that genes involved in energy homeostasis are more sensitive to food intake**. The DOHaD concept (Sect. 5.3) indicates that in times of continued famine, this epigenetic sensitizing is a survival advantage, while in times of plenty food it may drive the individual into obesity and T2D. DNA methylation is particularly sensitive to events during development *in utero*, because genomic DNA gets almost fully demethylated in the days

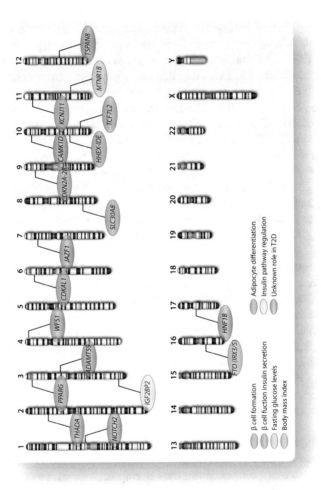

Fig. 9.8 Insights into the genetic basis of T2D. Examples of 18 genes are shown that were identified by GWAS to be associated with T2D. Only four genes were previously known as T2D candidate genes. The genes that participate in β cell disturbance have diverse functions, such as pancreatic islet proliferation, insulin secretion and cell signaling. The functional role of eight genes in T2D has not yet been identified

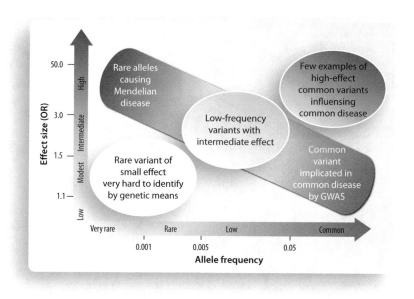

Fig. 9.9 Identifying genetic variants by risk allele frequency. The strength of the genetic effect is indicated by odds ratios. Most emphasis and interest lies in identifying associations with characteristics shown within the diagonal box

after zygote formation and specific methylation is re-established throughout embryogenesis. For example, patterns of DNA methylation at candidate genes, such as the *LEP* gene, associate with *in utero* exposure to famine and with maternal impaired glucose tolerance during pregnancy.

Epigenetic programming happens not only during the prenatal phase, but occurs also in the post-natal and adult phases of life. For example, T2D patients maintaining intense glucose control remain at increased risk of macrovascular complications and organ damage even years after the initial diagnosis. This "glycemic memory" is an epigenetic effect, *i.e.*, histone methylation changes within human aortic endothelial cells in response to increased glucose exposure. Individuals that carry an epigenome, which during their anthropologic development was programed by suboptimal nutrition *in utero*, despite normal post-natal nutrition, transgenerationally transmit a predisposition for obesity (Fig. 9.10b).

It is not clear how many generations epigenetic marks can be inherited, but it is obvious that epigenetic inheritance is not comparably persistent as genetic inheritance. Within more or less one generation (between 1981 and 2014) the worldwide prevalence for both obesity and T2D doubled. Thus, **populations that made a transition from famine to food surplus just within 1–2 generations are under significantly higher risk for obesity, T2D and the metabolic syndrome**, than those that were improving their nutritional conditions over many generations.

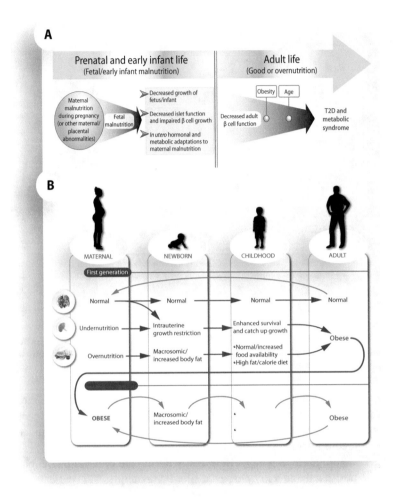

Fig. 9.10 Transgenerational view on T2D and obesity. T2D and the metabolic syndrome (Chap. 10) can be the outcome of maternal-fetal malnutrition, such as poor maternal diet, low maternal fat stores or reduced transfer of nutrients because of placental abnormalities (**a**). The fetus adapts to this environment by being nutritionally thrifty, resulting in decreased fetal growth, islet function and β cell mass and other hormonal and metabolic adaptations. A transition to overnutrition later in adult life exposes the impaired islet function to increased metabolic stress, which is further enhanced by obesity and age, so that T2D results. Non-obese mothers usually give birth to non-obese children, which develop into adults with a normal metabolic profile and a normal body fat content (**b**). However, undernutrition combined with improved neo-natal survival, formula feeding and exposure to a Western post-natal diet increases the incidence of prematurity and intrauterine growth restriction. This results in increased obesity of the offspring and higher risk for developing the metabolic syndrome, when prenatally exposed to Western diet. Some obese mothers may give birth to newborns with increased body fat, as a result of consumption of a high-fat diet. All these processes contribute to a shift of the population toward an obese phenotype. This also includes that second generation obese women have an increased risk to give birth to infants with increased body fat content and a further increased risk to develop obesity and the metabolic syndrome

For example, Polynesians show an overproportional high prevalence for T2D. One possible explanation is that their ancestors experienced a number of challenges, such as cold stress and starvation. This happened a few hundreds years ago during long open-ocean travels over the Pacific, which may have let only the initially most obese members of the group survive. These ancestors may have been evolutionary selected for energetic efficiency, referred to as thrifty metabolism. Accordingly, present-day Polynesians have inherited an increased T2D susceptibility, because their ancestors went through an evolutionary bottleneck. In contrast, populations, which did not experience long periods of starvation in the past centuries, such as the Europeans, have a low prevalence for T2D. Similar explanation may apply to other indigenous populations, such as the Pima Indian tribe in Arizona who had adapted to life of deprivation in the desert. When exposed to Western diet both Polynesians and Pima Indians get far more likely obese than Europeans. Thus, **populations who are born and live in countries that had particularly rapid changes in urbanization and economic development have an increased risk of the different features of the metabolic syndrome**.

Diabetes is a very heterogenous disease that in future will be diagnosed and treated on the basis of its molecular features, *i.e.*, on a far more personalized basis. One particular goal will be to use personalized lifestyle and medication, in order to reduce the risk of cardiovascular complications of T2D. Since most cases of T2D are preventable, in future far more effort will be taken in detecting early genetic and epigenetic markers for increased T2D susceptibility. Such epigenetic markers will help to detect children or families, for whom **intensive lifestyle intervention are likely to prevent the onset of metabolic disease**.

Additional Readings

Ashrafzadeh S, Hamdy O (2019) Patient-driven diabetes care of the future in the technology era. Cell Metab 29:564–575

Flannick J, Johansson S, Njolstad PR (2016) Common and rare forms of diabetes mellitus: towards a continuum of diabetes subtypes. Nat Rev Endocrinol 12:394–406

Ling C, Ronn T (2019) Epigenetics in human obesity and type 2 diabetes. Cell Metab 29:1028–1044

Perry RJ, Samuel VT, Petersen KF, Shulman GI (2014) The role of hepatic lipids in hepatic insulin resistance and type 2 diabetes. Nature 510:84–91

Stenvers DJ, Scheer F, Schrauwen P, la Fleur SE, Kalsbeek A (2019) Circadian clocks and insulin resistance. Nat Rev Endocrinol 15:75–89

Wells JCK (2017) Body composition and susceptibility to type 2 diabetes: an evolutionary perspective. Eur J Clin Nutr 71:881–889

Zimmet P, Shi Z, El-Osta A, Ji L (2018) Epidemic T2DM, early development and epigenetics: implications of the Chinese famine. Nat Rev Endocrinol 14:738–746

Chapter 10
Heart Disease and the Metabolic Syndrome

Abstract In this chapter, we will link three important risk factors for heart disease: hypertension, atherosclerosis and dyslipidemia. Chronically elevated blood pressure, *i.e.*, hypertension, increases the risk of ischemic heart disease, stroke, peripheral vascular disease and other CVDs. **This is the most important preventable risk factor for premature death**. The low intake of fruit and vegetables and whole grain fiber and high consumption of saturated fat and high-cholesterol diets can lead to hypercholesterolemia and atherosclerosis, especially in genetically predisposed individuals. Atherosclerosis is a chronic inflammatory disease caused by the accumulation of cholesterol-laden macrophages in the artery wall, *i.e.*, it is based on dyslipidemia and an overreaction of the immune system. Accordingly, the susceptibility to CVDs is associated with genes affecting the serum levels of plasma lipids and lipoproteins. Furthermore, we will discuss the role of insulin resistance and obesity in the major metabolic tissues liver, skeletal muscle, pancreas and WAT causing the metabolic syndrome. The importance of inflammation and regulation of energy metabolism will be highlighted. The genetic risk for the metabolic syndrome overlaps with that of its major components obesity, T2D and dyslipidemia. However, like in these traits, common genetic variations can explain only a minor part of the disease risk. Therefore, we will present **the important role of epigenetics in the origin and development of the metabolic syndrome**.

Keywords Hypertension · Atherosclerosis · Chronic inflammation · Foam cells · Cholesterol · Lipoproteins · LDL · HDL · Apolipoproteins · Dyslipidemia · Obesity · Insulin resistance · Metabolic syndrome · Liver · Adipose tissue · Skeletal muscle · Pancreas · Macrophages · Inflammation · Susceptibility genes · Epigenetics

10.1 Hypertension

Each cycle of heart contraction pumps some 70 ml blood into the systemic arterial system, in order to supply all cells and tissues of our body with oxygen and nutrients. This pulsation creates pressure on the vascular walls that depends on the amount of pumped blood and the resistance created by the vasculature. Due to the periodic ejection of blood from the heart, this pressure is highest at the peak of a passing amount of blood (systolic) and lowest after its passage (diastolic). Blood pressure displays a circadian rhythm with highest values in the afternoon and lowest at night. For healthy adults blood pressure values should be in the order 120 mm Hg (millimeters of mercury) systolic and 80 mm Hg diastolic, respectively (Fig. 10.1). Blood pressure is tightly regulated by signals from the SNS, in order to permit uninterrupted blood perfusion of all vital organs. For example, even **transient interruption in blood flow to the brain causes loss of consciousness and longer interruptions will result in death of non-perfused tissues, such as in cerebral stroke or myocardial infarction**.

Hypertension is defined as the blood pressure level above which therapeutic intervention has clinical benefit. Chronic hypertension in combination with atherosclerosis (Sect. 10.2) is the major risk factor for stroke, CHD, congestive heart failure and end-stage renal disease (Fig. 10.1). Obesity increases the risk of hypertension five-fold as compared with normal weight. Accordingly, more than 85% of hypertension cases are attributed to a BMI > 25. In 90–95% of all cases, hypertension results from a complex interaction of genes and environmental factors. GWAS identified more than 30 common SNPs with small effects on blood pressure. In addition, there are some rare genetic variants with large effects, which converge on a common pathway that alters blood pressure by changing the net renal salt balance. This emphasizes salt homeostasis in the kidney as a key risk factor for hypertension. Since humans originated from a notoriously salt-poor environment of sub-Saharan Africa (Sect. 1.1), gene variants that promoted salt and water retention provided a strong adaptive advantage. Nowadays, our body's biochemistry is still not fully adapted to the rather high salt load of our diet. Therefore, after a salty meal increased salt reabsorption in the kidneys requires water reabsorption, in order to maintain plasma sodium concentration at 140 mM. This results in an increased intravascular volume boosting venous blood return to the heart. Thus, **dietary factors significantly influence blood pressure: reduced dietary salt intake as well as increased consumption of fruits and low fat food, exercise, weight loss and reduced alcohol intake can reduce hypertension** (Sect. 1.4).

10.2   Mechanisms of Atherosclerosis

The endothelium is a single layer of endothelial cells that covers blood vessels and serves as a barrier between the circulating blood and subendothelial tissues. In atherosclerosis, cholesterol deposition below the endothelium causes a macrophage-dominated inflammatory response in large and medium arteries. **Atherosclerotic plaques tend to accumulate at the inner curvatures and branch points of arteries**, *i.e.*, at positions where laminar flow is either disturbed or insufficient, in order to maintain the normal, quiescent state of the endothelium. Some atherosclerotic lesions can develop already in the first years of life and 95% of humans by the age of 40 have some type of lesion. However, in most cases clinical manifestations do not occur before the age of 50–60 years.

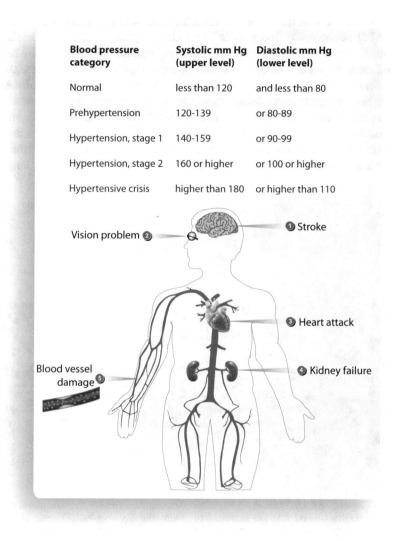

Blood pressure category	Systolic mm Hg (upper level)	Diastolic mm Hg (lower level)
Normal	less than 120	and less than 80
Prehypertension	120-139	or 80-89
Hypertension, stage 1	140-159	or 90-99
Hypertension, stage 2	160 or higher	or 100 or higher
Hypertensive crisis	higher than 180	or higher than 110

Fig. 10.1 Hypertension and its complications. Systolic blood pressure indicates how much pressure blood is exerting against the artery walls while the heart is pumping blood. The diastolic blood pressure measures the pressure while the heart is resting between beats. The ranges for normal blood pressure, prehypertension, hypertension (stages I and II) and hypertensive crisis are defined. Hypertension increases the risk of ischemic heart disease, strokes, peripheral vascular disease and other CVDs, including heart failure, aortic aneurysms, diffuse atherosclerosis and pulmonary embolism. It is also a risk factor for cognitive impairment and dementia as well as for chronic kidney disease. Other complications include hypertensive retinopathy and hypertensive nephropathy

A first sign of lesion formation is the accumulation of cholesterol within the arterial wall, referred to as fatty streaks. This is initiated by LDLs, which are cholesterol-rich, APOB-containing lipoproteins (Sect. 10.3). When LDLs are modified by oxidation, enzymatic and non-enzymatic cleavage and/or aggregation, they become pro-inflammatory and stimulate endothelial cells to produce chemokines, such as CCL5 and CXCL1, glycosaminoglycans and P-selectin for the recruitment of monocytes. Hypercholesterolemia (Sect. 10.3) and cholesterol accumulation in hematopoietic stem cells promote the overproduction of monocytes and their increased adherences to endothelial cells (Fig. 10.2). The monocytes then move into the space below the endothelial cells, referred to as "intima", where they differentiate into macrophages. These macrophages ingest normal and modified lipoproteins,

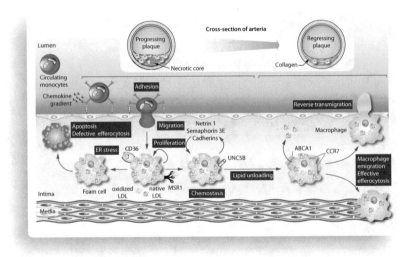

Fig. 10.2 Monocyte recruitment and accumulation in plaques. Hyperlipidemia increases the number of monocyte subsets that are recruited to atherosclerotic plaques. Different chemokine-chemokine receptor pairs and endothelial adhesion molecules mediate the infiltration of the monocytes into the intima. There the monocytes differentiate into macrophages or DCs and interact with atherogenic lipoproteins. Macrophages take up native and oxidized LDLs *via* macropinocytosis or scavenger receptors, such as MSR1 (macrophage scavenger receptor 1) and CD36, resulting in the formation of foam cells (Box 10.1). Imbalances in the lipid metabolism of macrophages within the progressing plaque can lead to the retention of the cells and subsequently to chronic inflammation. Retention molecules, such as netrin 1 and its receptor UNC5B (unc-5 homolog B), semaphorin 3E and cadherins, are expressed by lipid-laden macrophages and promote their chemostasis. Lipid unloading *via* ABCA1 and cholesterol efflux can reverse foam cell accumulation. In parallel, in myeloid-derived cells the chemokine receptor CCR7 is upregulated and the expression of retention factors is decreased

such as oxidized LDL, which at the onset of the inflammatory response is a beneficial clearance. In addition to hypercholesterolemia, also immunological and
mechanical injuries, as well as bacterial and viral infections, contribute to the
pathogenesis of atherosclerosis *via* the recruitment of macrophages. When the
inflammation turns chronic, the macrophages transform into cholesterol-laden foam
cells (Box 10.1). **These foam cells persist in plaques, *i.e.*, they have lost their
ability to migrate and cannot resolve inflammation**.

Box 10.1 Foam cells

One of the earliest pathogenic events in the nascent atherosclerotic plaque is
lipoprotein uptake by monocyte-derived macrophages. This results in the
development of foam cells (Fig. 10.3). Increased oxidative stress in the artery
wall promotes modifications of LDLs, which are primarily oxidations that are
recognized by macrophages *via* scavenger receptors, such as MSR1, CD36
and ORL1 (oxidized low density lipoprotein receptor 1). These receptors
internalize the lipoproteins and cholesteryl esters of the lipoproteins are
hydrolyzed to free cholesterol and fatty acids. Importantly, the scavenger
receptors are not downregulated by cholesterol *via* a negative-feedback mechanism, such as in the case of LDLR. Moreover, *via* macropinocytosis and
phagocytosis of aggregated LDLs, macrophages can internalize native LDLs
and VLDLs as well as oxidized lipoproteins. The internalized lipoproteins
and their associated lipids are digested in the lysosome releasing free cholesterol. The free cholesterol is transported to the ER, where it is re-esterified by
the enzyme ACAT1 (acetyl-CoA acetyltransferase 1) providing the "foam" in
foam cells. Enrichment of ER membranes with free cholesterol can result in
defective cholesterol esterification by ACAT1 in macrophages, promoting the
accumulation of free cholesterol. Lipophagy represents the delivery of lipid
droplets to lysosomes for efflux, while lipolysis by NCEH1 (neutral cholesterol ester hydrolase 1) mobilizes these lipids. The nuclear receptor LXR is
activated by the accumulation of cellular cholesterol (Sect. 3.4) and upregulates the expression of the transporter proteins ABCA1 and ABCG1 leading
the reverse cholesterol transport (Sect. 7.3). These proteins mediate the transfer of free cholesterol to lipid-poor APOA1, in order to form nascent HDLs or
mature HDLs, in which free cholesterol is esterified and stored in the core of
the particle. Cholesterol crystal formation in the lysosome is stimulated by
excessive free cholesterol accumulation and activates the NLRP3 inflammasome (Sect. 7.1), promotes ER stress and leads to apoptosis. Moreover, lipid
rafts that are enriched in sphingomyelin form a complex with free cholesterol
and promote TLR4 signaling. This leads to the activation of NFκB and the
production of pro-inflammatory cytokines and chemokines.

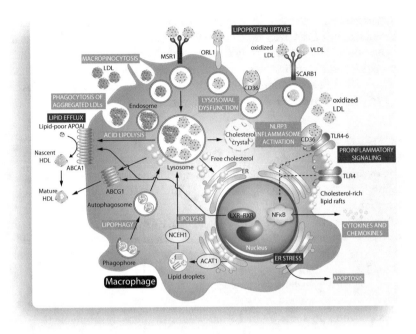

Fig. 10.3 Lipoprotein uptake and efflux in macrophages. Details are provided in the text

Many different cell types, such as endothelial cells, monocytes, DCs, lymphocytes, eosinophils, mast cells and smooth muscle cells, contribute to the process of atherosclerotic plaque formation, but foam cells are central to the pathophysiology of the disease. The TLR-dependent activation of these monocyte-derived cells polarizes them to M1-type macrophages (Sect. 7.1) that secrete proatherosclerotic cytokines, such as IL6 and IL12, matrix-degrading proteases as well as ROS, such as NO radicals. The plaque macrophages are subject to both retention and emigration signals and a misbalance of these processes contributes to the net accumulation of plaque macrophages. Dysregulation of lipid metabolism in foam cells contributes to ER stress ultimately resulting in apoptotic cell death. Since defective lipid metabolism pathways, such as cholesterol esterification and efflux, impair efficient clearance of apoptotic cells, **this results in necrosis and in the release of cellular components and lipids that form the necrotic core**.

Chronic inflammation stimulates the migration of smooth muscle cells into the intima region, where they transform into fibroblasts that proliferate and produce larger amounts of extracellular matrix. This leads to the formation of fibrous atherosclerotic plaques. Due to the calcification of the plaques, the artery wall becomes rigid, *i.e.*, sclerotic and fragile. Most humans have small atherosclerotic lesions that do not compromise blood flow. However, when the lesions grow and inward arterial remodeling, they gradually narrow the diameter of the blood vessels and the blood flow is decreased (referred to as stenosis). When this stenosis applies to more than 80% of the coronary arteries, the heart muscle becomes ischemic, especially when a high cardiac workload increases oxygen demand. Finally, **the originally stable lesion changes into an unstable vulnerable plaque that can easily rupture the endothelium leading to the formation of intravascular blood clots that can cause myocardial infarction or, in the case of damage of brain supplying arteries, in cerebral stroke**.

10.3 Lipoproteins and Dyslipidemias

Cholesterol is essential for membrane structure and fluidity and is a precursor to steroid hormones, vitamin D_3, oxysterols and bile acids that are ligands of nuclear receptors (Sect. 3.2). Only a small amount of circulating cholesterol originates from nutrition, while approximately 80% derive from endogenous synthesis. Cholesterol levels are tightly regulated by the coordinated actions of the transcription factors SREBF1 (Sect. 3.1) and LXR (Sect. 3.4). At low cholesterol levels, SREBF1 activates genes that are involved in endogenous cholesterol production and cholesterol uptake, such as *LDLR*. Since already low concentrations of free cholesterol can be toxic, cholesterol is mostly esterified with fatty acids. Cholesterol and cholesteryl esters are insoluble in plasma, which requires their transport in spheroidal macromolecules, referred to as lipoproteins. These lipoproteins have a hydrophobic core formed by phospholipids, fat-soluble anti-oxidants, vitamins and cholesteryl esters and a hydrophilic coat that contains free cholesterol, phospholipids and apolipoproteins.

There are four main types of lipoproteins: chylomicrons, VLDLs, LDLs and HDLs, which are differentiated based on their density and size (Box 10.2). The density of lipoproteins depends on the abundance of apolipoproteins. Chylomicrons and VLDLs are important for triacylglycerol transport and delivering fatty acids to peripheral tissues, while LDLs deposit cholesterol in peripheral tissues. **Increased levels of cholesterol-rich LDLs are associated with elevated risk of CVDs.** LDLs carry preferentially APOB and can deliver cholesterol to artery walls leading to the formation of atherosclerotic plaques (Sect. 10.2). HDLs have a high amount of APOA1 and mediate the reverse cholesterol transport to the liver (Sect. 7.3). Thus, **in contrast to LDLs, high levels of HDLs are associated with reduced risk for CVD**. The ratio of APOB to APOA1 is the strongest predictor of CHD risk, but the ratio of total cholesterol to HDL-cholesterol is equally predictive to this trait. For example, an increase of total cholesterol of 5.2–6.2 mM is associated with a three-fold elevated risk of death from heart attack, while a HDL-cholesterol level lower than 0.9 mM is the most common lipid disturbance of patients below the age of 60.

The main lipids in lipoproteins are free and esterified cholesterol and triacylglycerols (Fig. 10.4a). In triacylglycerol metabolism, hydrolyzed dietary fats enter intestinal cells *via* fatty acid transporters. Through a vesicular pathway, MTTP

Box 10.2 Lipoproteins
The composition of the four types of lipoproteins is listed.

Chylomicrons: With 50–200 nm diameter they represent the largest lipoproteins and show low density (< 1.006 g/ml). They are composed of approx.
85% triacylglycerols, 9% phospholipids, 4% cholesterol and 2% protein,
such as APOB48.

VLDLs: Very low density (0.95–1.006 g/ml) lipoproteins of 30–70 nm diameter containing approx. 50% triacylglycerols, 20% cholesterol, 20% phospholipids and 10% protein, such as APOB100.

LDLs: Low density (1.016–1.063 g/ml) lipoproteins of 20–25 nm diameter
that are composed of approx. 45% cholesterol, 20% phospholipids, 10%
triacylglycerols and 25% protein, such as APOB.

HDLs: High density (1.063–1.210 g/ml) lipoproteins with a diameter of
8–11 nm that are formed of approx. 40–55% protein, such as APOA1, 25%
phospholipids, 15% cholesterol and 5% triacylglycerols.

(microsomal triglycerole transfer protein) packs reconstituted triacylglycerols with
cholesteryl esters and APOB48 into chylomicrons. Chylomicrons also contain the
apolipoproteins APOA5, APOC2 and APOC3. In adipocytes, the enzyme DGAT1
(diacylglycerol O-acyltransferase 1) re-synthesizes triacylglycerols that had been
hydrolyzed by PNPLA2 and LIPE (hormone sensitive lipase). Chylomicron remnants are taken up by hepatic LDLR or LRP1 (LDLR-related protein 1). In liver
cells, triacylglycerols are packaged with cholesterol and APOB100 into VLDLs.
The triacylglycerols in VLDLs are hydrolyzed releasing fatty acids and VLDL remnants that are hydrolyzed by LIPC yielding LDLs. Sterols in the intestinal lumen
enter intestinal cells *via* the transporter NPC1L1 (Niemann-Pick C1-like protein 1)
and some are re-secreted by ABCG5 and ABCG8. In intestinal cells, dietary cholesterol is packaged with triacylglycerols into chylomicrons. In hepatocytes, cholesterol is recycled or synthesized *de novo* by a pathway, in which HMGCR is rate
limiting. LDLs transport cholesterol from the liver to the periphery, where they are
endocytosed. In HDL-cholesterol metabolism, APOA1 in HDLs mediates reverse
cholesterol transport by interacting with ABCA1 and ABCG1 transporters on non-
hepatic cells. LCAT esterifies cholesterol for the use in HDL-cholesterol, which is
remodeled by CETP and LIPG (endothelial lipase), in order to enter hepatocytes.

   In contrast to most other chronic inflammatory diseases, there is the potential to
remove the inflammatory stimulus in atherosclerosis. Lowering of plasma LDL levels by drugs, such as statins that target HMGCR can prevent subendothelial retention of lipoproteins and thereby decrease inflammatory atherosclerotic disease. The
enzyme PCSK9 (proprotein convertase subtilisin/kexin type 9) binds to LDLR and
induces its degradation, which reduces the metabolism of LDL-cholesterol and can
lead to hypercholesterolemia (Fig. 10.4a). The inverse correlation between HDL
levels (causing increased triacylglycerol levels) and the risk of CHD is due to the

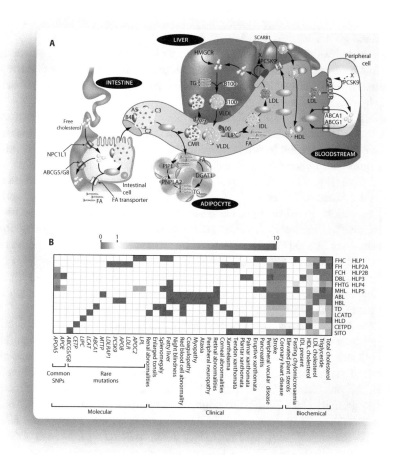

Fig. 10.4 Lipid metabolism and description of selected dyslipidemias. Details of lipid metabolism are described in the text (**a**). Dyslipidemias and their defining features are listed in rows and columns, respectively (**b**). The color intensity indicates for biochemical traits and for susceptibility to CHD, stroke and peripheral vascular disease (white: no difference from normal; red: a fold-increase above normal; blue: a fold-decrease below normal), for qualitative clinical features (white: absence of feature, *i.e.*, normal state; red: presence of the feature), for rare mutations (white: no role; red: a major etiologic role for the gene; pink: a minor etiologic role) and for common polymorphisms (white: no role; gradations of red: risk associated with the genotype). *CETPD* CETP deficiency, *FCH* familial combined hyperlipidemia, *FH* familial hypercholesterolemia, *FHC* familial hyperchylomicronemia, *FHTG* familial hypertriglyceridemia, HBL hypobetalipoproteinemia, *HLD* hepatic lipase deficiency, *HLP* hyperlipoproteinemia, *LCATD* LCAT deficiency, *MHL* mixed hyperlipidemia, *SITO* sitosterolemia

importance of HDL-cholesterol in reverse cholesterol transport from the periphery to the liver (Sect. 7.3). The association between HDL-cholesterol and CHD is complex, since HDLs contain a variety of proteins that are implicated in a number of biological pathways, such as oxidation, inflammation, hemostasis and innate immunity. This heterogeneity in the biological function of HDLs suggests that **the measurement of HDL-cholesterol levels alone is insufficient for reflecting the role of HDLs in atherosclerosis**.

Levels of certain plasma lipids and lipoproteins are key risk factors for CVD. Some 10% of the hypercholesterolemia cases have a monogenic basis, such as heterozygous familial hypercholesterolemia, familial defective APOB and autosomal dominant hypercholesterolemia that is based on a gain of function of the *PCSK9* gene (Fig. 10.4b). In contrast, different homozygous loss-of-function mutations in the genes *APOB* or *PCSK9* cause homozygous HBL (hypobetalipoproteinemia), in which almost no LDL-cholesterol is present. Homozygous mutations in *MTTP* cause a similar disease called ABL (abetalipoproteinemia). Rare monogenic disorders, such as TD (Tangier disease), affect HDL levels and are based on homozygous mutations in the *ABCA1* gene or deficiencies in the genes *APOA1*, *LCAT*, *CETP* or *LIPC*. Moreover, there are also rare monogenic disorders causing severe HTG (hypertriglycerolemia) that are due to homozygous loss-of-function mutations of the genes *LPL*, *APOC2* and *APOA5*.

Like in the case of monogenetic forms of obesity (Sect. 8.5) and T2D (Sect. 9.4), the identification of the genes causing monogenic dyslipoproteinemias significantly increased the understanding of the disease. For example, plasma LDL-cholesterol levels depend crucially on LDLR function, which in turn requires proper binding of APOB, the presence of LDLRAP1 (LDLR accessory protein 1) and intracellular LDLR degradation by PCSK9. The majority of hypercholesterolemia cases is based on common variants in genes, such as *APOE*, *LDLR*, *APOB*, *PCSK9* and *HMGCR* for LDL-cholesterol and *CETP*, *LIPC*, *LPL*, *ABCA1*, *LIPG* and *LCAT* for HDL-cholesterol, that display low ORs or epigenetic programming (Sect. 5.4). The genetic variants were identified either by SNP arrays or by targeted re-sequencing (Fig. 10.5). For example, an excess of mutations in the *NPC1L1* gene causes low intestinal sterol absorption or heterozygous mutations in the *LIPG* gene lead to high HDL levels.

The variations in lipoprotein levels that are based on common genetic variants are often too small to be meaningful in the clinical practice. However, these insights have a high value in basic research for the identification of novel pathways. Moreover, (epi)genetic profiling of individuals for metabolic diseases, such as dyslipidemias, allows a genetic risk stratification far earlier than the onset of the metabolic syndrome (Sect. 10.4). Such personalized medicine approaches will provide for the respective individuals longer and more effective periods of lifestyle changes. Since diet is a key determinant of lipoprotein levels, **early dietary interventions are the most efficient and most economic strategies for CVD prevention**. Furthermore, the results of biochemical assays, such as measurements of lipoprotein serum levels, should be integrated from multiple time points of the patient's lifetime.

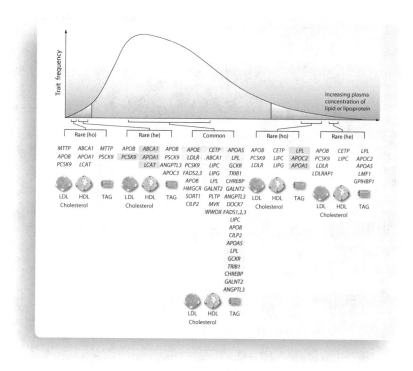

Fig. 10.5 Genetic variants affecting plasma lipoprotein distribution. The frequency of the traits LDL-cholesterol, HDL-cholesterol and triacylglycerol levels (y axis) is plotted over plasma concentrations (x axis). The bottom and top fifth percentiles of the distribution are indicated by shaded areas. The genes (no shading: by classical genetic or biochemical methods; orange: by resequencing; blue: by GWAS) that determine lipoprotein concentrations in specific segments of the distribution are shown below the respective graphs. The extremes of the distribution represent homozygotic (ho) monogenic disorders, the less extremes heterozygous (he) mutations and the center common variants. The green shading indicates small to moderate effect sizes associated with severe HTG

10.4 Whole body's Perspective of the Metabolic Syndrome

Nowadays the metabolic syndrome is a very common aging-related condition that occurs primarily as a result of overweight and obesity caused by a sedentary lifestyle, *i.e.*, physical inactivity, and the consumption of diet containing excess of calories (Sect. 8.4). The syndrome is composed of different factors that either alone or in combination significantly increase the risk of T2D and CVDs. Most of these risk factors have been discussed in the previous chapters, such as visceral obesity (Sect. 8.1), ectopic lipid overload (Sect. 9.2), insulin resistance (Sect. 9.2), β cell failure (Sect. 9.3), hypertension (Sect. 10.1) and dyslipidemias with high plasma concen-

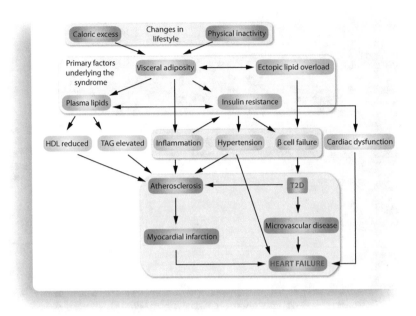

Fig. 10.6 Interactions of metabolic syndrome traits. Changes in lifestyle, such as increased consumption of high-caloric diets combined with reduced physical activity, play important roles in the worldwide dramatic increase in the metabolic syndrome. Visceral obesity, ectopic lipid overload, dyslipidemias and insulin resistance are the primary factors underlying the syndrome. These factors cause inflammation, hypertension and β cell failure. Individuals with the metabolic syndrome are therefore at increased risk for the development of atherosclerosis, T2D, microvascular diseases, myocardial infarction and finally heart failure

trations of triacylglycerols and low concentrations of HDL-cholesterol (Sect. 10.3) (Fig. 10.6). **The dramatic worldwide increase in obesity and the parallel rise in life expectancy, *i.e.*, the increased number of elderly around the world, make the metabolic syndrome a major global health problem**.

Historically, the concept of "syndrome X" was used to describe the metabolic syndrome in the end of 1980s as a condition with increased risk of T2D and CVD caused by insulin resistance in metabolic tissues. Since then the *National Cholesterol Education Program* (*NCEP*), the *WHO*, the *European Group for the study of Insulin Resistance* (*EGIR*) and the *International Diabetes Federation* (*IDF*) used slightly different thresholds to define the metabolic syndrome based on rate of obesity, hyperglycemia, dyslipidemias and hypertension (Table 10.1). While the *NCEP* definition does not require any defined parameter, the *WHO* proposes that an evidence of insulin resistance, such as impaired glucose tolerance, impaired fasting glucose or T2D, is essential. In contrast, the *EGIR* emphasizes hyperinsulinemia as main

Table 10.1 Definitions of the metabolic syndrome

	NCEP (2005)	WHO (1998)	EGIR (1999)	IDF (2005)
Absolutely required	None	Insulin resistance	Hyperinsulinemia	Central obesity
Criteria	Any of the 5 below	Insulin resistance or T2D	Hyperinsulinemia	Obesity
		Plus 2 of 5 below	Plus 2 of the 4 below	Plus 2 of the 4 below
Obesity	Waist circumference:	Waist/hip ratio:	Waist circumference:	Central obesity
	Males > 101.6 cm	Males >0.90	Males > 94 cm	
	Females > 88.9 cm	Females >0.85 Or BMI > 30 kg/cm^2	Females > 80 cm	
Hyperglycemia	Fasting glucose > 5.6 mM	Insulin resistance	Insulin resistance	Fasting glucose > 5.6 mM
Dyslipidemia I	Triacylglycerols > 1.7 mM	Triacylglycerols > 1.7 mM Or HDL < 0.9 mM	Triacylglycerols > 2.0 mM Or HDL < 1.0 mM	Triacylglycerols > 1.7 mM
Dyslipidemia II	HDL: Males < 1.0 mM Females <1.25 mM			HDL: Males < 1.0 mM Females <1.25 mM
Hypertension	> 130 mm hg systolic or > 85 mm hg diastolic	> 140/90 mm hg	> 140/90 mm hg	> 130 mm hg systolic or > 85 mm hg diastolic
Other criteria		Microalbuminuria		

criterium, while for the IDF central obesity is essential. At present, the definitions of *NCEP* and *IDF* are most widely used.

Our body has integrated mechanisms to become either catabolic, when energy demands cannot be met by food intake, or anabolic, when calorie supply exceeds energy demands. The key regulator of these mechanisms is insulin, which is secreted by β cells in the pancreas after a meal and promotes carbohydrate resorption, energy utilization (*via* glycolysis), storage of carbohydrates (as glycogen), storage of fat (as triacylglycerols) and synthesis of fat from carbohydrates (*via* activating *de novo* lipogenesis) in key metabolic tissues. In parallel, insulin inhibits lipolysis, *i.e.*, the release of energy from triacylglycerols, and the synthesis of glucose (*via* gluconeogenesis) after a meal. Thus, **the actions of insulin create an integrated set of signals that represent the nutrient availability and the energy demands of our body**. In turn, a disturbance in insulin actions, such as obesity-triggered insulin resistance in one or multiple metabolic organs, such as skeletal muscle, liver and WAT, often serves as the onset of the metabolic syndrome (Fig. 10.7). These conditions can lead to organ-specific consequences, such as β

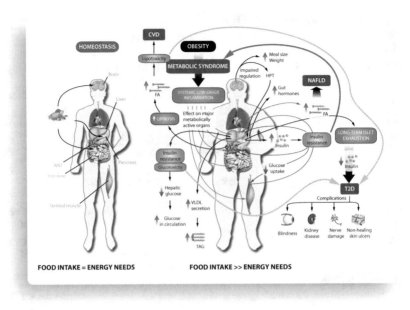

Fig. 10.7 Whole body's view on the metabolic syndrome. Under normal conditions, energy intake and utilization is perfectly balanced with our body's energy needs (**left**). Intestine, pancreas and brain sense food after a meal and send signals to muscle, liver, fat and back to the brain, in order to maintain metabolic homeostasis *via* the coordination of uptake and storage of nutrients and energy production. The metabolic syndrome often starts with obesity, triggering a state of systemic low-grade inflammation that affects various major organs involved in metabolic homeostasis (**right**). The brain's ability to regulate meal size or frequency is impaired leading to weight gain and further organ dysfunction. The autonomic nervous system and the hypothalamic-pituitary-thyroid (HPT) axis are disrupted causing a change in the release of gut hormones. Insulin resistance is a further important trigger of the metabolic syndrome. In the pancreas, the islets expand to release more insulin (hyperinsulinemia) in an attempt to overcome insulin resistance of muscle, liver and WAT. However, over time the islets become exhausted and little or no insulin is produced, so that T2D occurs. Insulin resistance in the muscle leads to excessive glucose uptake in the liver that is primarily converted to fatty acids often causing NAFLD. Moreover, in the liver, glucotoxicity and insulin resistance result in inefficient downregulation of hepatic glucose production leading to further increase of circulating glucose levels. The fat excess in the liver can be released into the circulation as VLDLs leading to elevated serum triacylglycerol levels. Insulin resistance of adipose tissue increases its lipolytic activity, thus also releasing excess fatty acids. All together these lipid sources result in lipotoxicity that further contributes to organ dysfunction and disease, especially CVD. Lipotoxicity and glucotoxicity worsen T2D and lead to numerous complications, such as kidney disease, blindness, nerve damage and non-healing skin ulcers

cell failure and NAFLD, but also to systemic effects, such as glucotoxicity, lipotoxicity and low-grade inflammation. All these conditions are key factors of the metabolic syndrome and accelerate the risk of diabetes, heart disease and their complications.

10.5   Metabolic Syndrome in Key Metabolic Organs

Systemic effects of the metabolic syndrome influence the metabolism in key metabolic organs, such as liver, muscle, pancreas and WAT. Hepatic insulin resistance causes elevated activity of the key gluconeogenesis enzymes G6PC and PCK, and increased glycogenolysis, both leading to increased glucose output from the liver (Fig. 10.8a). In parallel, the expression of enzymes that regulate glycogen synthesis and glycolysis, such as GCK and pyruvate kinase, is reduced, driving GLUT2 to transport glucose out of hepatocytes. All these alterations in the glucose metabolism pathways accelerate systemic glucotoxicity. In addition, a decreased insulin sensitivity in the liver increases the uptake of FFAs and the formation of triacylglycerols. These are loaded to VLDLs for transport in the circulation and thus cause dyslipidemia. The increase of glucose levels in the liver increases lipogenesis *via* the activity of SREBF1 and FASN, which are both not impaired by insulin resistance. This accumulation of lipids in liver can cause NAFLD. Hepatic insulin resistance also contributes to hyperlipidemia through the downregulation of LDLR (Sect. 10.3). In this way, liver insulin resistance causes decreased clearance of LDLs and VLDLs, leading to increased LDL and VLDL levels in the circulation, respectively.

Skeletal muscle is the main tissue for glucose storage and utilization. After a meal approximately 80% of the glucose load of the blood is taken up by the muscle. Insulin resistance in muscles leads to reduced insulin-stimulated GLUT4-mediated glucose transport into the myocytes (Fig. 10.8b). Decreased glucose uptake reduces the levels of glucose-6-phosphate to be used for glycogen synthesis and glycolysis. This increases the glucose concentration in the bloodstream and causes systemic glucotoxicity. Like the liver, systemic lipotoxicity also overloads the muscle with FFAs. These lipids are taken up and stored in form of triacylglycerols in intramuscular lipid droplets.

In the early stages of insulin resistance, β cells increase the production and secretion of insulin, in order to maintain glucose tolerance (Fig. 10.9a). Since under these conditions insulin is less potent in suppressing hepatic glucose production, the liver becomes insulin resistant. When insulin resistance progresses, β cells lose their ability to compensate for decrease insulin response *via* an increase of insulin release. This finally results in reduced circulating insulin concentrations and often comes along with increased glucagon levels. This shift in the glucagon-insulin ratio leads to a further rise in hepatic gluconeogenesis and advanced hyperglycemia occurs. Systemic glucotoxicity and lipotoxicity, *i.e.*, constant exposure of β cells to elevated levels of glucose and lipids, both increase glucose metabolism in β cells and cause metabolic stress leading to the unfolding protein response of the ER in these cells. In response to ER stress, hypoxic stress and pro-inflammatory cytokines, the β cells fail to proliferate and undergo uncontrolled autophagy or even apoptosis. This leads to β cell dysfunction and ultimately their death.

In obesity, the storage capacity of adipocytes is often exceeded causing cellular dysfunctions, such as increased formation of ceramides, ER stress and hypoxia leading to reduced metabolic control and cell death. An increase in number and size

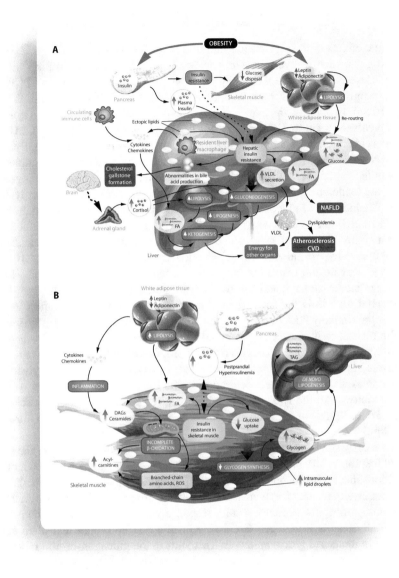

Fig. 10.8 Metabolic syndrome in the liver and skeletal muscle. Early stages of the metabolic syndrome are associated with insulin resistance causing a decreased glucose disposal in skeletal muscle. Obesity also alters the secretion pattern of adipokines, such as increased levels of leptin and decreased adiponectin concentrations. Together this leads to a re-routing of glucose and lipids (as a consequence of increased lipolysis from adipose tissue) to the liver (**a**). As intracellular lipid levels in the liver rise, along with the increased plasma insulin, hepatic insulin signaling rapidly deteriorates and this impairment turns the liver into a "co-conspirator" in the further progression of the metabolic syndrome and its complications. Hepatic insulin resistance leads to increased hepatic glucose production. Further hormones can contribute to decreased insulin sensitivity of the

of adipocytes also influences the secretion of adipokines. In addition, adipocytes attract monocytes into WAT that become M1-type macrophages, secreting pro-inflammatory cytokines, together with adipokines leading to low-grade systemic inflammation. All this contributes to insulin resistance in WAT causing a reduced insulin-stimulated import of glucose *via* GLUT4 (Fig. 10.9b). Moreover, lipolysis is increased due to the impaired inhibition of LIPE activity leading to an increase in FFA release from adipocytes. The ability of insulin to stimulate the re-esterification of FFAs is also impaired, and consecutively systemic lipotoxicity occurs.

The metabolic and inflammatory response of metabolic tissues, such as WAT, integrates the actions of the innate immune system, *i.e.*, primarily of the monocyte-derived macrophages, with those of adipocytes (Fig. 10.9b). This integrated action of two or more different tissues has developed during evolution. Responses of the immune system to pathogen invasions use substantial amounts of energy for new protein synthesis and rapid growth of cells. Thus, under these conditions **it makes sense that metabolic organs are insulin insensitive, *i.e.*, that they use less energy from circulating glucose and lipids during limited time periods**. For this reason, inflammatory mediators control energy metabolism so that pathogens are defended most efficiently, *e.g.*, through the ability to shift rapidly from glucose oxidation to lipid oxidation. With a similar protective attempt lipids can trigger insulin resistance, in order to preserve glucose for glucose-dependent organs, such as the CNS and erythrocytes.

10.6   Genetics and Epigenetics of the Metabolic Syndrome

GWAS analysis for central factors of the metabolic syndrome, such as BMI, T2D and dyslipidemia, have identified for each of these traits highly statistically significant associations with 40 to 100 genetic variants (Sects. 8.5, 9.6 and 10.3). The key genes of these lists, such as *LPL*, *APOE*, *MC4R*, *FTO* and *TCF7L2* (Table 10.2), are

---

Fig. 10.8 (continued) liver. For example, cortisol increases glucose production, promotes lipolysis and increases lipid deposition. Insulin resistance increases the secretion of VLDLs from the liver and causes dyslipidemia. Excessive hepatic secretion of VLDLs plays an important role in promoting atherosclerosis and CVDs. Insulin resistance also increases intrahepatic fat accumulation finally causing to NAFLD. Activation of inflammatory pathways occurs both systemically, as a result of cytokines released from circulating immune cells, and locally through resident liver macrophages. This further accelerates lipid accumulation and storage. Liver insulin resistance also causes abnormalities in bile acid production and increases the risk for cholesterol gallstone formation. Insulin resistance in skeletal muscle causes reduced insulin-stimulated glucose uptake and therefore less glucose is available for insulin-stimulated glycogen synthesis (**b**). Overload of lipids increases the level of intramyocellular lipids in the form of DAG and ceramides as well as increases in acyl-carnitines (due to incomplete mitochondrial fatty acid β-oxidation). Pro-inflammatory adipokines, branched-chain amino acids and ROS further contribute to this defect in insulin signaling. Insulin resistance in skeletal muscle promotes post-prandial hyperinsulinemia and diversion of ingested carbohydrates away from storage as glycogen in skeletal muscle to liver where they are converted to triacylglycerols through increased hepatic *de novo* lipogenesis

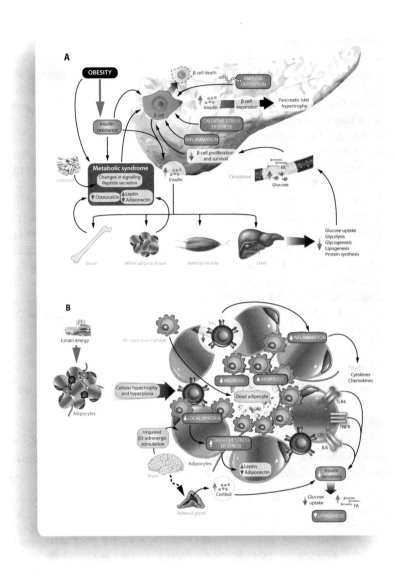

Fig. 10.9 Metabolic syndrome in the pancreas and adipose tissue. Obesity promotes the development of insulin resistance leading to compensatory increases in insulin release from β cells (**a**). Chronic overproduction of insulin leads to β cell expansion, *i.e.*, pancreatic islet hypertrophy. As insulin resistance progresses, the effects of insulin on target tissues diminish, thereby leading to impairments in glucose uptake, glycolysis, glycogenesis, lipogenesis and protein synthesis. This leads to hyperglycemia and elevated FFA levels in the circulation that negatively affect β cell proliferation and survival. This results in a vicious cycle that further reduces β cell function. In the setting of the metabolic syndrome, many tissues show alterations in hormone levels that directly

also the central determinants for the genetic risk for the metabolic syndrome. However, the dilemma with all these common SNPs remains that their individual ORs are clearly below 2, *i.e.*, they contribute considerably less than 100% increased risk for the disease. This implies that the common SNPs can explain only a small fraction of the cases of the metabolic syndrome. This suggests that **socio-environmental factors and epigenetic mechanisms rather than variations of the core genome play a role in the obesity epidemic and its associated metabolic abnormalities**.

Integrative genomic approaches can combine large-scale genome- and transcriptome-wide data, in order to construct gene networks that underlie metabolic traits, such as plasma lipoproteins levels. For example, plasma HDL-cholesterol levels were linked to variants in the regulatory region of the *VNN1* (vanin 1) gene using RNA expression profiles in lymphocytes. This type of analysis indicates that metabolic traits are the products of molecular networks being modulated by sets of complex genetic loci and environmental factors. This re-emphasizes that **the genetic predisposition for a metabolic disease, such as atherosclerosis, is comprised of multiple common genetic variants that each have a small to moderate effect on the trait, either alone or in combination with modifier genes or environmental factors**.

There are more and more epidemiological and clinical evidences that the DOHaD concept (Sects. 5.4 and 9.6), *i.e.*, a prenatal epigenetic programming *in utero*, may be the key cause for the metabolic syndrome. So far, there is no comprehensive analysis of the epigenome of persons suffering from the metabolic syndrome, *i.e.*, no concrete genomic regions of elevated risks have been identified. However, it can be assumed that due to the complexity of insulin signaling and its interference with multiple pathways (Sect. 6.3) a high number of regions will be affected in an individual-specific way. Nevertheless, since epigenetic modifications respond dynamically to environmental conditions (Chap. 5), there is potential for intervention and reversibility.

Fig. 10.9 (continued) impact β cell function. The intestine changes the secretion of signaling peptides, the bone secretes less osteocalcin and WAT secretes more leptin but less adiponectin. These hormonal changes together with oxidative stress, ER stress, inflammation and intracellular amyloid deposition cause β cell death. In the presence of excess energy supply WAT expands as a result of cellular hypertrophy and hyperplasia (**b**). These enlarged adipocytes become dysregulated through an increased rate of local necrosis, apoptosis and pro-inflammatory responses. Dead adipocytes attract macrophages that are conventionally skewed towards an M1-like pro-inflammatory profile. Under obese conditions there is also a reduction of T_{REG} cells. This causes an increase in the local pro-inflammatory environment that can ultimately spill over to systemic increases in pro-inflammatory cytokines. The rapid tissue expansion during obesity leads to local hypoxia and the activation of ER stress response causing a reduced release of insulin-sensitizing adipokines, such as adiponectin. Moreover, increased levels of cortisol and the activation of TLRs and other pro-inflammatory cytokine receptors, such as TNFR and IL6 receptors, further reduce insulin sensitivity. This leads to reduced rates of triacylglycerol synthesis, increasing levels of FFAs and a decrease in insulin-mediated glucose uptake. In contrast, the impaired β3-adrenergic response downstream of SNS activity leads to reduced metabolic flexibility, since FFAs cannot be appropriately activated in response to β3-adrenergic stimulation

Table 10.2 Central genes in the development of metabolic syndrome

Gene locus	Gene function	Disease context	Affected parameter
LPL	Hydrolyzes triacylglycerols (Sect. 3.4)	Cardiovascular	HDL concentration
APOE	Removing lipoproteins from circulation (Sect. 10.3)	Cardiovascular	HDL concentration
MC4R	Membrane receptor on neurons binding α-MSH (Sect. 8.4)	Obesity	Waist circumference
FTO	Function mediated by IRX3 and IRX5 (Sect. 8.7)	Obesity	Waist circumference
TCF7L2	Transcription factor in β cells (Sect. 9.6)	T2D	Glucose concentration

Healthy dietary patterns, such as Mediterranean or Nordic diet (Box 1.1), lower the risk of the metabolic syndrome. **Studies understanding the molecular mechanism of diet on epigenetic programming in the prenatal, post-natal and adult phases of life are of major importance, in order to understand how diet can prevent the metabolic syndrome.** Thus, cleverly designed dietary intervention studies and observational studies will investigate the impact of individual nutrients, such as vitamin D_3 or PUFAs, within a healthy dietary pattern, in order to improve the conditions of the metabolic syndrome. In these studies, a larger number of epigenome-, transcriptome-, proteome- and metabolome-wide data will need to be integrated.

Additional Readings

Barroso I, McCarthy MI (2019) The genetic basis of metabolic disease. Cell 177:146–161

Drummond GR, Vinh A, Guzik TJ, Sobey CG (2019) Immune mechanisms of hypertension. Nat Rev Immunol 19:517–532

Hunter PM, Hegele RA (2017) Functional foods and dietary supplements for the management of dyslipidaemia. Nat Rev Endocrinol 13:278–288

Jensen MK, Bertoia ML, Cahill LE, Agarwal I, Rimm EB, Mukamal KJ (2014) Novel metabolic biomarkers of cardiovascular disease. Nat Rev Endocrinol (11):659–672

Tall AR, Yvan-Charvet L (2015) Cholesterol, inflammation and innate immunity. Nat Rev Immunol 15:104–116

Glossary

Acute inflammation a short-term immunological process occurring in response to tissue injury or microbe infection usually appearing within minutes or hours. It is characterized by pain, redness, swelling and heat.

Adaptive introgression movement of genetic material from the genome of one species into the genome of another through repeated interbreeding.

Admixture interbreeding between previously isolated populations.

Adipogenesis the process whereby fibroblast-like progenitor cells restrict their fate to the adipogenic lineage, accumulate nutrients and become triglyceride-filled mature adipocytes.

Adipokines cytokines secreted by adipose tissue.

Adiponectin a peptide hormone produced in adipose tissue that is involved in regulating glucose levels as well as fatty acid breakdown.

Anabolism the building-up aspect of metabolism, in which a set of metabolic pathways construct molecules from smaller units by the use of energy.

Anatomically modern humans individuals classified as *Homo sapiens* on the basis of a set of morphological characteristics that distinguish them from other, now extinct, archaic humans. According to the fossil record they emerged approximately 300,000 years ago.

Atherosclerosis a disease in which the inside of an artery narrows due to the build up of plaque. It can result in coronary artery disease, stroke, peripheral artery disease or kidney problems, depending on which arteries are affected.

Atherosclerotic plaques are the underlying entities of atherosclerotic diseases.

Autophagy a natural, regulated mechanism cells that removes unnecessary or dysfunctional components.

β-oxidation the catabolic process by which fatty acid molecules are broken down in the mitochondria in eukaryotes, in order to generate acetyl-CoA.

Bioactive compound a type of chemical found in small amounts in plants and certain foods, such as fruits, vegetables, nuts, oils and whole grains promoting good health.

© Springer Nature Switzerland AG 2020
C. Carlberg et al., *Nutrigenomics: How Science Works*,
https://doi.org/10.1007/978-3-030-36948-4

Bromodomain a protein module of ~110 amino acids that mediates interaction with acetylated lysines and is often found in HATs and ATP-dependent chromatin remodeling proteins.

Brown adipose tissue (BAT) Highly specialized adipose tissue, the main function of which is to produce heat (thermogenesis).

Calorie restriction a dietary regimen that reduces calorie intake daily without incurring malnutrition or a reduction in essential nutrients.

Cancer a group of diseases involving abnormal cell growth with the potential to invade or spread to other parts of the body.

Cardiovascular disease (CVD) a class of diseases that involve the heart or blood vessels, such as coronary artery diseases, stroke, heart failure and more.

Catabolism the set of metabolic pathways breaking down molecules into smaller units that are either oxidized to release energy or used in anabolic reactions.

Chromatin the molecular substance of chromosomes being a complex of genomic DNA and histone proteins.

Chromatin immunoprecipitation followed by sequencing (ChIP-seq) a method for genome-wide mapping of the distribution of histone modifications and chromatin associated proteins that relies on immunoprecipitation with antibodies to modified histones or other chromatin proteins. The enriched DNA is sequenced to create genome-wide profiles.

Chromatin modifying enzymes/chromatin modifiers proteins that are either recognizing (reading) chromatin (*i.e.*, post-translationally modified histones and methylated genomic DNA), adding (writing) marks or removing (erasing) them.

Chromodomain a modular methyl-binding domain of 40–50 amino acids that is commonly found in chromatin remodeling proteins.

Chronic inflammation long-term inflammation lasting for prolonged periods of several months to years. Chronic inflammation plays a central role in most common non-communicable diseases, such as cancer, T2D, asthma and Alzheimer's.

Circadian clock a biochemical oscillator that cycles with a stable phase and is synchronized with solar time.

CpG CG dinucleotides (the "p" indicates the phosphate linking the two nucleosides). Out of 16 possible dinucleotides, CpGs are the only ones that can be methylated symmetrically; *i.e.*, DNA methylation can be inherited only *via* CpGs to both daughter cells.

Cytokines a category of small proteins (~5–20 kDa) that are important in cell signaling involved in autocrine, paracrine and endocrine signaling as immunomodulating agents.

Damage-associated molecular patterns (DAMPs) also known as alarmins, are molecules often released by stressed cells undergoing necrosis that act as endogenous danger signals to promote and exacerbate inflammatory responses. Examples of non-protein DAMPs include cholesterol crystals and SFAs. DAMPs are associated with many inflammatory diseases, including arthritis, atherosclerosis, Crohn's disease and cancer.

Diploid an organism or cell with a double set of chromosomes, so that each position is represented by two genes or alleles.

DNA methylation the covalent addition of a methyl group to the C5 position of cytosine.

DNA methyltransferases (DNMTs) family of enzymes catalyzing the transfer of a methyl group to cytosines of genomic DNA.

Developmental Origins of Health and Disease (DOHaD) an approach to investigate the role of prenatal and perinatal exposure to environmental factors, such as undernutrition, in determining the development of human diseases in adulthood.

Dyslipidemia an abnormal amount of lipids, such as triglycerides, cholesterol and/ or fat phospholipids, in the blood.

Embryogenesis the process by which an embryo forms and develops. In mammals, the term is use exclusively to the early stages of prenatal development, whereas the terms fetus and fetal development describe later stages.

Energy balance a measurement of the biological homeostasis of energy in living systems.

Enhancer a stretch of genomic sequence that (like a promoter) contains clusters of transcription factor binding sites that regulate a gene within the same TAD.

Epigenetic drift a divergence of the epigenome as a function of age due to stochastic changes in DNA methylation or stable histone modifications.

Epigenetic epidemiology the study of the relationship between epigenetic variants and disease phenotype in the population.

Epigenetic memory a heritable change in gene expression that is induced by a previous developmental or environmental stimulus. It requires chromatin-based changes, such as DNA methylation, histone modifications or incorporation of variant histones.

Epigenetic programming the process leading to stable and long-lasting alterations of the epigenome based on specific covalent modifications of the DNA and histones.

Epigenetics the study of heritable changes in gene function that do not involve changes in the DNA sequence. Epigenetic mechanisms include the covalent modifications of DNA and histones.

Epigenome the complete set of epigenetic modifications across an individual's genome.

Epigenomics studies of the epigenome.

Expression quantitative trait locus (eQTL) genomic loci that explain all or a fraction of variation in expression levels of mRNAs.

Euchromatin light-staining, decondensed and transcriptionally accessible regions of the genome.

Evolution change in the heritable characteristics of biological populations over successive generations.

Evolutionary pressure any cause that reduces reproductive success in a portion of a population driving natural selection.

Gene expression process by which information from a gene is used in the synthesis of a functional gene product. These products are often proteins, but can also be ncRNAs.

Genetic architecture the landscape of genetic contributions to a given phenotype. It comprises the number of genetic variants that influence a phenotype, the size of their effects on the phenotype, the frequency of those variants in the population and their interactions with each other and the environment.

Genetic drift changes in genetic variation over time that are due to random (by chance) processes, *i.e.*, different from natural selection in evolution.

Gene regulatory networks represent units of interacting proteins that are functionally constrained by defined regulatory relationships. These interactions provide a structure and determine an output in the form of a pattern of gene expression. Networks are usually visualized by nodes (proteins) and edges (their interactions).

Genome the complete haploid DNA sequence of an organism comprising all coding genes and far larger non-coding regions. The genome of all 400 tissues and cell types of an individual is identical and constant over time (with the exception of cancer cells).

Genome-wide association study (GWAS) studies that aim to identify genetic loci (mostly SNPs) associated with an observable trait, disease or condition.

Genotype the genetic makeup of an organism, *i.e.*, its complete heritable genetic identity.

Glucagon a peptide hormone produced by α cells of the pancreas raising the concentration of glucose and fatty acids in the bloodstream.

Gluconeogenesis a metabolic pathway for the synthesis of glucose from precursor substrates, such as lactate and amino acids.

Glucotoxicity the toxic effects of excessive levels of glucose in the blood (as in T2D).

Glycemic index a number from 0 to 100 (for pure glucose) representing the relative rise in the blood glucose level 2 h after consumption.

Glycogenolysis the breakdown of glycogen to glucose-1-phosphate.

Glycogenesis the process of glycogen synthesis, in which glucose molecules are added to chains of glycogen for storage.

Glycolysis the metabolic pathway converting glucose into pyruvate.

Haploid a single set of chromosomes.

Haplotype block a genomic region with no evidence of a history of genetic recombination.

Healthspan the duration of disease-free physiological health within the lifespan of an individual. In human, for instance, this corresponds to the period of high cognitive abilities, immune competence and peak physical condition.

Hematopoiesis the formation of blood cellular components derived from HSCs.

Hematopoietic stem cells (HSCs) stem cells located in the bone marrow that can develop into all types of blood cells.

Heritability the proportion of total variation between individuals within a population that is due to genetic factors.

Heterochromatin dark-staining, condensed and gene-poor regions of the genome.

High-density lipoprotein (HDL) an 8–11 nm high-density (1.063–1.210 g/ml) lipoprotein with 40–55% protein (with APOA1 as the major apolipoprotein),

25% phospholipids, 15% cholesterol and 5% triglycerides. HDL particles carry cholesterol from peripheral tissues to the liver.

Histone code an epigenetic code that is based on post-translational modifications of histone proteins. The histone modifications serve to recruit other proteins by specific recognition of the modified histone *via* specialized protein domains. The code comprises more than 130 post-translational modifications serving as an "alphabet" for the instructions, how the epigenome directs transcriptional regulation and stores information.

Homeostasis the state of steady internal physical and chemical conditions maintained by living systems.

Hyperglycemia permanently elevated blood glucose levels as a result of an insufficient use or production of insulin.

Hyperplasia formation of new cells through differentiation of resident precursors.

Hypertension a long-term medical condition, in which the aterial blood pressure is persistently elevated.

Hypertrophy increase in size of existing cells.

Impaired glucose tolerance blood glucose levels that are raised beyond normal levels (7.8–11.0 mM 2 h after a 75 g OGTT) but not high enough to warrant a diabetes diagnosis.

Imprinted genes represent more than 100 human genes that are expressed in a parent-or-origin manner, i.e., this epigenetic phenomenon is independent of classical Mendelian inheritance.

Inflammaging a chronic low-grade inflammation developing with advanced age and accelerating the process of biological aging and worsening many age-related diseases.

Inflammasome a supramolecular complex responsible for the CASP1- dependent maturation of IL1B and IL18 in response to microbial products or other danger signals.

Insulin a peptide hormone produced by β cells of the pancreatic islets serving as the main anabolic hormone of the body.

Insulin resistance a pathological condition, in which the systemic and cellular response to insulin action is impaired.

Integrative personal omics profiling (iPOP) an analysis method that combines genomic, transcriptomic, proteomic, metabolomic and autoantibody profiles from individuals in a longitudinal over a period of multiple months to years.

Lactase persistence the continued activity of the lactase enzyme in adulthood.

Linkage analysis an approach attempting to identify regions in chromosomes that co-segregate with the trait of interest in related individuals.

Linkage disequilibrium an association between two alleles that are located so close to each other on the genome that they are inherited together more frequently than expected by chance.

Lipogenesis a metabolic pathway for the synthesis of fatty acids and triglycerides.

Lipolysis the metabolic pathway through which lipid triglycerides are hydrolyzed into glycerol and three fatty acids.

Lipoproteins a vesicle whose primary purpose is to transport lipids in water, such as in blood plasma or other extracellular fluids.

Lipotoxicity a cellular dysfunction arising from accumulation of lipid intermediates in cells other than adipocytes.

Low-density lipoprotein (LDL) a 20–25 nm low-density (1.016–1.063 g/ml) lipoprotein with ~45% cholesterol, 20% phospholipids, 10% triglycerides and 25% protein (with APOB as the major apolipoprotein).

Macrophage a type of white blood cell engulfs and digesting cellular debris, foreign substances, microbes and cancer cells in a process called phagocytosis.

Massive parallel sequencing also called next-generation sequencing (NGS), high-throughput approach to DNA sequencing using the concept of massively parallel processing.

Metabolic syndrome a cluster of conditions, such as increased blood pressure, high blood sugar, excess body fat around the waist and abnormal cholesterol or triglyceride levels, that occur together, increasing the risk of heart disease, stroke and T2D.

Metaflammation low levels of inflammation throughout the body due to excessive nutrient consumption.

Meiotic recombination the reciprocal physical exchange of chromosomal DNA between the parental chromosomes occurring at meiosis during spermatogenesis and oogenesis.

Missing heritability the fact that genetic variations cannot account for all of the heritability of diseases, behaviors and other phenotypes.

Monocytes leukocytes of the innate immune system that can differentiate into macrophages and myeloid lineage DCs.

Myocardial infarction also known as a heart attack occurring when blood flow decreases or stops to a part of the heart, causing damage to the heart muscle.

Myokines small proteins or proteoglycans that are released by myocytes upon contraction to mediate autocrine, paracrine or endocrine effects.

Neutrophils the most abundant type of granulocytes and the most abundant type of white blood cells belonging to the innate immune system.

Non-alcoholic fatty liver disease (NAFLD) A very common disease in humans in which there is an excessive accumulation of fat in the liver (steatosis) in individuals who are not alcoholic.

Non-coding RNA (ncRNA) an RNA molecule that is not translated into a protein.

Non-communicable disease a disease that is not transmissible directly from one person to another, such as autoimmune diseases, CDVs, most cancers, diabetes and Alzheimer's disease.

Nuclear receptor a transcription factor that can be activated by a small lipophilic ligand in the size of cholesterol.

Nucleosome a basic unit of DNA packaging in eukaryotes, consisting of 147 bp of genomic DNA wound around a histone octamer.

Nutrigenomics a discipline studying the relationship between human genome, nutrition and health.

Obesity a medical condition in which excess body fat has accumulated to an extent that it may have a negative effect on health.

Odds ratio (OR) the mathematical expression of the relation between the presence or absence of a variant, *e.g.*, a SNP, and the presence or absence of a trait, *e.g.*, a disease, in the population.

Oxidative phosphorylation the metabolic pathway in which cells use enzymes to oxidize nutrients, thereby releasing energy, which is used to produce ATP.

Pattern recognition receptors evolutionarily conserved receptors that elicit inflammation and innate immunity upon recognition of conserved microbial products or endogenous danger signals, such as DAMPs.

Pathogen-associated molecular patterns (PAMPs) small molecular motifs derived from microbes, such a lipopolysaccharides. They are recognized by toll-like receptors and other pattern recognition receptors on the surface of cells of the innate immune system.

Personalized nutrition a conceptual analog to personalized medicine, where individuals are recommended to take certain food products based on nutrigenomics approaches.

Phenotype primarily physical, externally visible traits of an organism, but also may include internal and microscopic or biochemical traits.

Polygenic risk scores a weighted sum of the number of risk alleles carried by an individual, where the risk alleles and their weights are defined by the loci and their measured effects as detected by genome wide association studies.

Post-translational modifications covalent modifications, such as phosphorylations, acetylations or methylations, by which most proteins reach their full functional profile. Due to post-translational modifications the proteome is far more complex than the transcriptome and also varies a lot in response to extra- and intracellular signals.

Promoter stretches of genomic DNA for productive transcription initiation encompassing at least one TSS.

Proteome in analogy to the transcriptome, the complete set of all expressed proteins in a given tissue of cell type. The proteome depends on the transcriptome, but is not its 1:1 translation.

Quantitative trait loci (QTLs) genomic regions at which genetic variation is associated with molecular variation across individuals. For example, individuals with a particular single nucleotide variant have altered expression levels of a gene (eQTL), altered DNA methylation (meQTL; also known as mQTL) or altered chromatin state (chromQTL).

RNA sequencing (RNA-seq) a method using massive parallel sequencing to reveal the presence and quantity of RNA in a biological sample at a given moment.

(Retro)transposon a transposon (also called transposable element or "jumping DNA") is a DNA sequence that can change its position within a genome. When this transposition is mediated *via* an RNA intermediate, the term retrotransposon is used.

Reverse cholesterol transport a multi-step process resulting in the net movement of cholesterol from peripheral tissues back to the liver first *via* entering the lymphatic system and the bloodstream.

Signal transduction pathway a process by which a chemical or physical signal is transmitted through a cell membrane as a series of molecular events, such as protein phosphorylation catalyzed by protein kinases. Mostly, signal transduction pathways end in the activation of a transcription factor or a chromatin modifier.

Single nucleotide polymorphism (SNP) a substitution of a single nucleotide at a specific position in the genome, which is present to some appreciable degree within a population (*e.g.*, more than 1%).

Sirtuins (SIRTs) a family of seven NAD^+-dependent HDACs that are structurally and mechanistically distinct from Zn^{2+}-dependent HDACs. Sirtuins influence a wide range of cellular processes such as aging, transcription, apoptosis, inflammation and stress resistance.

Stenosis is an abnormal narrowing in a blood vessel.

Stroke a medical condition in which poor blood flow to the brain results in cell death.

Thermogenesis a process by which cells generate heat.

Toll-like receptor (TLR) a class of PPRs that play a key role in the innate immune system.

Trait a distinguishing quality or characteristic belonging to a person.

Transcription factors proteins that sequence-specifically bind to genomic DNA. Our genome encodes approximately 1600 transcription factors, referred to as *trans*-acting factors, since they are not encoded by the same genomic regions, which they are controlling. Accordingly, the process of transcriptional regulation by transcription factors is often called *trans*-activation.

Transcription start sites (TSSs) nucleotides within a promoter that are the first to be transcribed by Pol II into a particular RNA.

Transcriptome the complete set of all transcribed RNA molecules of a tissue or cell type. It significantly differs between tissues and depends on extra- and intracellular signals.

Transgenerational epigenetic inheritance transmission of epigenetic information that is passed on to gametes without alteration of the DNA sequence.

Transrepression a process whereby one protein represses (*i.e.*, inhibits) the activity of a second protein through a protein-protein interaction.

Tricarboxylic acid (TCA) cycle a series of chemical reactions used by all aerobic organisms to release stored energy through the oxidation of acetyl-CoA derived from carbohydrates, fats, and proteins.

Type 2 diabetes (T2D) a form of diabetes being characterized by high serum glucose levels, insulin resistance and relative lack of insulin.

Unfolded protein response a cellular stress response of the ER.

Variant a difference from the reference or standard sequence, *i.e.*, a polymorphic site, including SNPs and structural deletions or insertions (indels). It can also encompass much larger chromosomal rearrangements (translocations, duplications, or deletions) that result in CNVs.

Western diet a dietary pattern characterized by high intakes of red meat, processed
   meat, prepackaged foods, butter, fried foods, high-fat dairy products, eggs,
   refined grains, potatoes, corn and high-sugar drinks.
White adipose tissue (WAT) the major type of adipose tissue primarily storing
   triglycerides in one large vacuole per cell.
Zoonotic pathogens pathogens naturally transmitted between animals and humans.

Printed in the United States
By Bookmasters